SpringerBriefs in Environment, Security, Development and Peace

Volume 21

Series editor

Hans Günter Brauch, Mosbach, Germany

More information about this series at http://www.springer.com/series/10357
http://www.afes-press-books.de/html/SpringerBriefs_ESDP.htm

Kyoko Kusakabe · Rajendra Shrestha
Veena N.
Editors

Gender and Land Tenure in the Context of Disaster in Asia

Springer

Editors
Kyoko Kusakabe
School of Environment Resources
and Development
Asian Institute of Technology
Klong Luang, Pathum Thani
Thailand

Veena N.
School of Environment Resources
and Development
Asian Institute of Technology
Klong Luang, Pathum Thani
Thailand

Rajendra Shrestha
School of Environment Resources
and Development
Asian Institute of Technology
Klong Luang, Pathum Thani
Thailand

The views expressed in this publication are those of The Author(s) and do not necessarily reflect the views or policies of the Food and Agriculture Organization of the United Nations.

ISSN 2193-3162 ISSN 2193-3170 (electronic)
SpringerBriefs in Environment, Security, Development and Peace
ISBN 978-3-319-16615-5 ISBN 978-3-319-16616-2 (eBook)
DOI 10.1007/978-3-319-16616-2

Library of Congress Control Number: 2015934435

Springer Cham Heidelberg New York Dordrecht London

Cover photo: A view of the field houses being used by indigenous people who face resettlement in Luang Namtha province of Lao PDR. This photo was taken by: Kyoko Kusakabe who granted the permission to use it here). More on this book is at: http://www.afes-press-books.de/html/SpringerBriefs_ESDP_21.htm

Copyediting: PD Dr. Hans Günter Brauch, AFES-PRESS e.V., Mosbach, Germany

Printed on acid-free paper

Springer International Publishing AG Switzerland is part of Springer Science+Business Media
(www.springer.com)

Acknowledgments

We would like to acknowledge the continuous support and cooperation of Yuji Niino, Land Management Officer, FAO-RAP, Bangkok for supporting the Expert Consultation on Land tenure and disaster and its social and gender impact organized by the Asian Institute of Technology in November 2013.[1] We would like to thank Prof. Sivanappan Kumar, VP (Academic Affairs), AIT, for his encouragement and support. We also extend our thanks to all the people at FAO and AIT who supported the expert consultation workshop from which this book is drawn. We acknowledge with gratitude the contribution of all participants at the workshop through their comments and suggestions. We thank our publishers Johanna Schwarz and our series editor Hans Günter Brauch for their immense patience and encouragement. Finally, we would like to extend our grateful thanks to reviewers Ben White, Clara Park, Dietrich Schmidt-Vogt, Donna L. Doane, Dung T. Ngo, Gayatri Lokuge, Govind Kelkar, Kanchana N. Ruwanpura, Keiko Ikeda, Mokbul Morshed Ahmad, Padmapani Perez, Ranjit Perera, Ranjith de Silva, Sharyn Graham, Smita Mishra Panda, Wang Yunxian and Zenaida Willison for their critique which was very useful in improving the chapters of this book.

Bangkok, December 2014

Kyoko Kusakabe
Rajendra Shrestha
Veena N.

[1]A short report on this Expert Consultation Workshop on Land tenure and disaster and its social and gender impact in Bangkok, 14–15 November 2013 is at: http://www.fao.org/fileadmin/templates/rap/files/meetings/2013/131209_report-land.pdf

Contents

Chapter 1
Gender and Land Tenure in the Context of Disaster

Veena N. and Kyoko Kusakabe

Abstract Disasters are both physical and social events and their impact is disproportionately borne by the most disadvantaged communities with the fewest resources to mitigate it. Women's reproductive and care responsibilities also increase the impact of disaster. Women also have fewer resources to rely on, which hampers their recovery process. In disaster situations, land rights form a key resource. The vulnerability of people, specifically women, with insecure land rights increases exponentially in a disaster situation. As a result, efforts are underway to formalise land rights of marginalized people and increase joint titling of marital property. The case studies in this book suggest that such well-meaning efforts do not always help women; on the contrary, they increase women's vulnerability to disaster by weakening and eroding their customary land rights. A nuanced approach to gender and land rights would be more useful to women and other disadvantaged groups in the process of disaster preparedness, mitigation and recovery and reconstruction.

Keywords Disaster · Land tenure · Gender roles · Land rights · Vulnerability

1.1 Introduction

A disaster impacts on people and environment at many levels—physical, social, economic, psychological, cultural and even spiritual. However, the extent of impact varies widely. The impacts of disasters are disproportionately borne by the most disadvantaged people and communities around the world. This is because (1) they have few resources and hence cannot afford to mitigate the impact of the disaster; (2) the poorest normally live in places at high risk of disaster; and (3) recovery is

---

Veena N. (India) is a researcher in the School of Environment Resources and Development at the Asian Institute of Technology, Klong Luang, Pathum Thani, Thailand, Email: veena.n@ait.ac.th; Kyoko Kusakabe (Japan/Thailand) is an Associate Prof. in the School of Environment Resources and Development at the Asian Institute of Technology, Klong Luang, Pathum Thani, Thailand, Email: kyokok@ait.ac.th.

K. Kusakabe et al. (eds.), *Gender and Land Tenure in the Context of Disaster in Asia*, SpringerBriefs in Environment, Security, Development and Peace, DOI 10.1007/978-3-319-16616-2_1

much harder because they have few or no resources to rebuild their lives. Gender plays a key role in placing women at a disadvantage due to their gender roles, gendered barriers to mobility, information, social capital and even food and nutrition.

Weaker claim to resources, especially land resources increases vulnerabilities and reduces resilience against risks. In most countries, women own less property than men; and have weaker land rights. Indigenous people's land rights, in particular, are threatened as governments and the private sector compete for forest land. The vulnerabilities of people with weak claims to land are well documented (Mehta 2007; Agarwal 1994). Further, we recognize that women's vulnerability does not stem from lack of access to land alone, but also from other gender-based deprivation and inequalities that multiply the effect of natural and human-made disasters. It is with this realization that we begin our book.

Gender discrimination in entitlement to land intersects with other inequalities and practices, and intensifies women's vulnerability to disaster. Therefore, unpacking women's vulnerabilities in a disaster situation is much more complicated than it would have been if the problem were merely about land rights and land titles. We problematize current patterns of land distribution and ownership in the disaster mitigation cycle which tend to further disadvantage the already disadvantaged—usually women.

The case studies in this book investigate efforts by NGOs and governments in Asia to help women, indigenous people and other disadvantaged groups affected by natural and human-made disasters. In each case, we find that their concerted effort to mitigate disaster risks or recover from disaster can, contrary to the intent, weaken women's claims to land. Our case studies suggest that a nuanced approach to gender and land rights would be more useful to the most disadvantaged groups in disaster preparedness as well as post-disaster recovery and reconstruction. We will first review some relevant literature on land and disaster, and gender issues in land tenure. Then, we reflect on the literature from the perspective of gender and disaster, and discuss how women's claims to land can be compromised when gender is conflated with other inequalities during a disaster. We then introduce the case studies in the book looking at women's land rights in the event of disasters, both natural and human-made, and draw hopes and strategies for the future.

1.2 Land and Disaster

Socio-economic changes caused by the twin imperatives of globalisation and modernization are drastically changing land use patterns all over Asia. International demand for land and natural resources has led to the conversion of farm and forest land into plantations and *export processing zones* (EPZ). Poor rainfall, rising ocean levels, droughts, flash floods, and other disasters are intensifying (Land Watch Asia 2012). To cope, farmers and disadvantaged groups often end up selling or leasing their land. Land is not just a productive asset and a source of material wealth, but equally a source of security, status and recognition (Rao 2011).

Disasters such as typhoons, tsunamis, volcanic eruptions, and earthquakes can erase land records thus dispossessing landowners (Land Watch Asia 2012), and leave vulnerable and disadvantaged groups open to land grabbing by powerful people. McClean (2010) notes that outright violence and carefully orchestrated legislative measures are often used to grab the land of the urban poor following disasters, accentuating inequalities. Poor urban governance and lack of accountability have increased the post-disaster risk of displacement in the Philippines (Ginnetti et al. 2013).

Although Asia is home to some of the fastest growing populations and markets in the world, much of the region's property is unregistered and held without legal title (see Chaps. 2–4). Informal property owners cannot gain access to formal bank loans and are subject to land-grabbing attempts (Fandl 2005). In the event of a disaster, survivors without formal property rights lose both homes and businesses and have no claim to the natural and financial resources necessary for rebuilding. Insecurity of tenure makes individuals vulnerable to exploitation in the aftermath of a disaster. Post disasters, the sooner people safely return to their land, the less dependent they are on humanitarian hand-outs (Brown/Crawford 2006).

When land tenure insecurity intersects with prevailing socio-economic, political and environmental challenges, the vulnerability of societies is accentuated, notes Chagutah (2013) following research in southern Africa. Handmer/Choong (2006) maintain that land grabbing increases post-disaster in areas where land tenure is informal, especially in remote coastal and forest areas. Such instances of people with weak official claims to land losing their land altogether was seen in post-tsunami Thailand. Abdulharis (2006) concluded that the Indonesian land administration system was not equipped to deal effectively with post-disaster rehabilitation and reconstruction needs of the people of Banda Aceh after land records and farm boundaries were wiped out. Marginal land owners found it extremely difficult to handle the bureaucratic processes involved in reclaiming their official land title. Similarly, Mitchell (2010) states that in the aftermath of a natural disaster, people lose access to land through resettlement and opportunistic land grabbing, which in turn leads to a loss of livelihood. He stresses on the importance of tenure security in post-disaster reconstruction and restitution.

Both *de jure* and *de facto* tenure security promote disaster resilience by providing secure access to land for housing and livelihood and provide a platform for disaster risk reduction and a better recovery program (Usamah 2013). Studying the relationship between land tenure security and disaster resilience in the Philippines, Usamah (2013) notes that perceived tenure security can provide opportunities for disaster risk reduction and promote disaster resilience. In a pre-disaster setting, tenure security provides access to housing and livelihood, which are pre-conditions for secure livelihoods. In a post-disaster setting, tenure security aids in early recovery and reconstruction. Land is fundamental for recovery from disasters as a site for shelter, a resource for livelihoods and a place to access services and infrastructure (Fricska et al. 2010).

As the demand for land grows around the world, disadvantaged groups are unable to defend their rights to land against corporate and powerful interests, leaving them more vulnerable in the face of disaster (see Chaps. 2, 5, and 7). Land as a resource is important for the disadvantaged to secure access to in order to strengthen their capability to cope with disaster. But their rights can be precarious during a disaster depending on other social and political structures. When other powerful people intend to grab land, there is little that the disadvantaged can do to defend themselves even when they had entitlement to land. This is why Wisner et al. (2004) note that it is not only the economic systems at national and international level but also the social systems and power relations that determine people's experience of disaster. These social systems include those that discriminate against women in terms of land ownership, inheritance and increased workload of women for unpaid work. In the next section, we review the linkage between gender and land tenure in Asia.

1.3 Gender and Land Tenure in Asia

Gender inequality in women's ownership over land and property in Asia is determined by various factors—economic, political, social and cultural. It is shaped by patriarchal inheritance practices that discriminate against women, restricted credit schemes that limit women's purchase of land, and social norms that oppose women's ownership of land, women's weak control and decision-making over their land, gender-biased or gender-neutral land reform (Kabeer 2009; Rao 2006, 2010).

Schlager and Ostrom (1992) categorise property rights into use rights and control rights. Use rights or usufruct rights guarantee access to land only as long as it is in use; lack of use results in lack of land rights. Such land rights are very weak and in a post-disaster situation, usufruct rights can be denied and land can be claimed by stronger social groups or individuals. Often, women have only usufruct rights to land and their access to land is indirect, i.e. mediated through their relationship with specific men (fathers or husbands), and their usufruct rights are an extension of their role to support the family, not as a recognition of their individual rights. Control rights to land include the right to manage the land, exclude others from the land, and alienate the land through sale, gift or inheritance. Such rights are often recorded in land management systems and are stronger than use rights. For example in Nepal, until recent legal reform, women were not able to inherit land from their parents, which put them in a weaker position in the household (CSRC, undated). Land rights is further more important because access to productive resources such as water, irrigation systems, and even capital are tied to land tenure in most countries (Meinzen-Dick et al. 1997).

When communal lands are taken over by the government and handed over to private interests, the entire community is deprived of access to land and other natural resources, as can be seen in the case of the Iban people of Sarawak in Chap. 2 of this book. In such instances, it is women in their gendered roles as mothers, wives, daughters and caretakers of the sick and young in the community who bear the brunt of the dispossession (see Yong and Wee; Herath; Jayatilaka; Yong/Siahaan/Burghofer, this volume). Women use forest land and other public land to get access to fodder, fuel, food, water and *non-timber forest products* (NTFPs). When the State hands over land rights to private entities, women's access to such forest land is restricted. For example, in Northern Laos, the expansion of economic concession to rubber plantation companies has deprived women of NTFPs they collected from forest (Kusakabe/Vongphakdy 2014; also see Raman; Yong/Siahaan/Burghofer, this volume).

On the other hand, joint titling or other forms of land titling do not in any way guarantee women's entitlement to land. In joint titles, the land is usually still controlled by men, and hence they do not always serve women's interests; rather joint titles keep women in abusive relationships (Razavi 2007). Such joint land titling programmes do not necessarily empower women to retain control over produce and income, nor to claim their share in case of marital conflict (Agarwal 2003) because the focus is on maintaining family harmony and economy rather than ensuring gender equity. Knight et al. (2012: 14) note: "community land documentation may be a more efficient method of land protection than individual or family land titling". Strengthening local systems for land governance is one of the processes recommended to strengthen women's land rights, maintain rights that may be lost from oral to written tradition, rejuvenate customary rights, and align local rules with national rules to protect women's rights.

Gender and land has been studied in great detail in Asia over the past few decades (Agarwal 1994; FAO 2005, 2007). Many factors influence women's rights to land including culture, customary laws and entitlements, tribal practices, legal systems, legal practices, Syariah law, etc. Across the developing world, these rights are changing in response to various pressures at the local and global level (Julia/White 2012; Scalise 2009; Villareal 2006; White et al. 2012; Wisborg 2014). In some countries, Syariah laws are incorporated into statutory law; while in others women-friendly customary laws such as *adat* are applied and legal norms are widely ignored (FAO 2007). Customary law in Asia is extremely diverse; in patrilineal Punjab women have no direct land rights whereas in matrilineal Thailand, women can inherit and sell property (FAO 2007). As Jackson (2003) argues, blanket prescriptions do not support the land rights of different subgroups of the population given the complexities of cultures, globalization and disaster.

1.4 Gender and Disaster

The gendered impacts of disaster have been researched in considerable depth over the past decade (Enarson/Chakraborti 2009; de Mel et al. 2009; Fitzpatrick 2008; Enarson/Morrow 1998; Mehta 2007). Women experience disaster differently from men, mainly because of their gender roles and responsibilities, as well as their gender-mediated access to information and other resources (Pincha 2008). Women, especially poor women, are more vulnerable to disaster, because of lower preparedness resulting from lack of information and resources. They tend to recover more slowly from disasters because they have a smaller buffer to rely on, and a much lower social capital to draw from when they need help.

Researchers note that one of the reasons why women experience disasters more intensely than men is because of their different levels of vulnerability. Women's vulnerabilities, including their weak claim over resources, specifically land, are directly linked to their socio-economic position, which is in turn mediated by gender within patriarchal structures (Agarwal 1994; Meinzen-Dick et al. 1997; Rao 2010). Women tend to have few resources to begin with, and in the event of disaster, these meagre resources are further depleted (see Jayatilaka, this volume). This makes women's recovery after disaster more difficult, exacerbating their vulnerabilities further. It also reduces their resilience to future shocks (see Herath; Jayatilaka, this volume).

Wisner et al. (2004: 5) note that disasters are a product of social, political and economic environments because of the way these structure the lives of different groups of people including "the manner in which assets, income and access to other resources, such as knowledge and information, are distributed between different social groups, and various forms of discrimination that occur in the allocation of welfare and social protection". As noted in the previous section, land is largely owned by men, and such unequal ownership of land leads to women being disproportionately affected by disaster (see Yonariza/Mahdi, this volume). This not only increases women's vulnerability to disaster, but also post-disaster.

It is not gender itself that marks vulnerability, "but gender *in a specific situation*" (Wisner et al. 2004: 15; *italics by authors*). Many vulnerability situations are temporary, and change as life stages do. Normal "systems of economic, social and political power contribute to creating vulnerability to disaster" (Wisner et al. 2004: 18). Using the access model, household livelihoods including social relations, structure of domination and other aspects of normal life become the starting point to map the different factors that play a role in the disaster process.

Such vulnerability is exacerbated because disaster relief organizations tend to see women as passive victims and ignore their needs to secure productive resources. Hewitt (1997: 167) note that treating socially vulnerable groups as 'special needs groups' tends to convert them into passive recipients, even victims. In the context of *disaster risk reduction* (DRR) in Liberia, Holmgren and Moe (2012) note that disaster relief organizations see women only as beneficiaries and not as contributors to change. A study of recovery from the floods that hit Queensland and Victoria in 2011 showed that women were considered victims and their potential to contribute significantly in regaining economic stability for themselves and their families was not fully recognized (Shaw et al. 2012). Women use their agency to creatively develop strategic coping mechanisms including sexual, familial and economic bargains to manage their livelihood options and resist the constraints of patriarchy (Kandiyoti 1988). A gender-blind response to natural disasters can reinforce, perpetuate, and increase existing gender inequality, making bad situations worse for women (APWLD 2005; see Jayatilaka, this volume).

1.5 Gender, Land and Disaster

As discussed, weak land tenure makes people more vulnerable to disaster. Women's weaker land entitlements vis-a-vis men make women more vulnerable than men. However, gender and disaster literature suggests that women's vulnerability under disaster is multiplied by other deprivations and inequities, as well as from gender-blind relief operations. Trohanis et al. (2011) identify four factors that put women at particular risk from natural disasters and have implications for how natural disaster risk management programs are designed and implemented. These include: (1) vulnerability; (2) livelihoods; (3) education and participation; and (4) land titling and inheritance rights. These are not independent variables, but are integrated with one another and map out the quality of life women and men lead. For example, unstable livelihoods or dependency in livelihoods leads to vulnerability as much as low education and weak land titles do. Therefore, during the pre-disaster stage, women are already more vulnerable than men, making their experience during disaster different from that of men. In the post-disaster period, women have more difficulty regaining their livelihoods and land, since their limited participation in decision making as well as pre-disaster vulnerability make their issues a lower priority in terms of emergency assistance and reconstruction. In addition, gender-blind government policy redefine women's roles as helpless victims and as wives of income earners, not recognizing or supporting women's agency.

While an insecure land title is an important risk factor determining the impact of disasters on women, this impact is greatly exacerbated by other factors. Women's general vulnerability—their dependency on men and other members of the household economically and socially—can weaken their claim in the household and society, and create a disadvantageous position for them. That is compounded by their weaker livelihood options because of their lack of access to productive resources including financial capital, lack of mobility to market their goods and lack of time to juggle various responsibilities in the household (see Jayatilaka, this volume). Women's low education actually leads to limited livelihood options and greater vulnerability, and lack of participation in decision making in the household and community which in turn means their voices are not heard and their needs are ignored. Women's vulnerability to disaster through their weak entitlement to land as discussed above is compounded by these other disadvantages.

The case studies in this book point to these entangled vulnerabilities. Many point out women's weak claim to land, and how it makes women more vulnerable. Disaster pushes women further into risk, but its impact is far greater than it should be, since women are also disadvantaged by other inequalities such as lack of decision making power and lack of access to financial resources, and are invisible as agents and active players to disaster relief agencies and governments. Such vulnerability is further aggravated by gender-blind relief operations by government and aid organizations.

We have brought together six case studies in this book to present the variety of ways in which land tenure impacts on women and disadvantaged people during disasters, and at the same time, how other factors—government policies and relief operations, social norms and practices—further complicate women's existing vulnerability. We hope these case studies will demonstrate the complicated web of vulnerability that affects women during a disaster, and how efforts by well-meaning outsiders can make things worse if local contexts are not taken into consideration.

1.6 Case Studies from Marginalized Communities in Asia

In their chapter titled 'Gender impact of large-scale deforestation and oil palm plantations on indigenous communities in Sarawak, Malaysia', *Carol Yong* and *Wee Aik Pang* note that for the indigenous Iban people of Sarawak, the development of oil palm plantations has been an unmitigated disaster. The plantations have destroyed the region's forests and people's culture, religion, lifestyle and livelihoods. Their access to water and food has been compromised, and women suffered disproportionately because of their responsibilities to secure food and water for the family. Indigenous communities are wreaked by divisions between those who support plantation development and those who are against. After a long struggle of 14 years, in which women also played significant role in supporting the struggle, a small part of the plantation has been given back to the community. However, rather than converting the plantation back to forest, community men leaders decided to keep it as palm oil plantation for cash income for the community. That is, women continue to be deprived of forest land even after their struggle to regain the land. With the entry of cash income and employment for men, women are losing the rights they had in the traditional longhouse. Thus, in the event of a disaster such as a flood, quake or drought, their vulnerabilities are very high and their capacity to recover from it is weakened further.

Subhangi Herath described how gender-blind policy, with all its good intentions, is creating disadvantages for women in their risk mitigation effort. In the chapter titled 'Displacing women, resettling families: Impact of landslides on women's land tenure rights in Sri Lanka', she notes that landslides occur with alarming regularity in the hilly areas of Sri Lanka. A study of a voluntary displacement and resettlement process in Nawalapitiya, a landslide-affected hilly town in Sri Lanka, showed that resettlement affected women more than men. Nawalapitiya's residents were told to move to a resettlement area since their houses were located in places with a high risk of landslide. Although many systems of marriage and inheritance are practiced in the area, officials in charge of land and resettlement tend to display a patriarchal bias in their work. Women's land, which has been inherited from their natal families, are registered in the name of the men, since they were the heads of households. Women and their natal families were complicit in this process as vocal opposition could displease husbands and break up families. The resettlement process thus ended up denying women their few rights and increasing their vulnerabilities. In the event of another disaster, women find themselves without even the few resources that they had earlier. Several women said that they preferred to live with the risks of landslides than to give up their land in favour of their husbands in the process of resettlement. They claimed that their mothers and grandmothers said the land was safe, while they know well that the place is unsafe. Women were vulnerable in both situations—by staying, they risked the landslide; by leaving, they risked losing land.

The next chapter by *Yonariza* and *Mahdi* also described how well-meaning government policies to recover land titles after flood disaster have deprived women of their traditional land entitlements. They studied land tenure change among the matrilineal Minangkabau of Sumatra, where all the agricultural land traditionally belongs to the clan and not the individual. It is inherited by the daughters, and guarded by their eldest brother. All clan members have equal access to land for subsistence purposes, though nobody can sell, lease or otherwise alienate the land. In their chapter titled 'Impact of flash floods on land ownership in a matrilineal society in West Sumatra, Indonesia', *Yonariza* and *Mahdi* analyse the post-flood damage assessment carried out by government officials and local leaders. They report that of the 155 instances of loss of land, livestock and trees, 102 instances were recorded as belonging to men, 52 to women, and only one instance of livestock loss was recorded to a group. With formal land titling, the formal 'owner' of the land was recorded as the *mamak* (maternal uncle), and official ownership shifted from the clan to the individual who was the nominal guardian of the land. Thus, land records developed by local leaders as well as government officials showed that most of the land officially belongs to men. These records are important, since they were used not only to decide who receives government compensation for the 2012 flood loss, but also for future developments on the land. Since the official list only names the 'owners' of the land, it does not reflect all users of the land—women, family members, lease-holders and share croppers. Thus, these groups do not receive any compensation for loss. This has far reaching implications for the future development of the Minangkabau community in Sumatra.

Not only does gender–blind policy for disaster mitigation and recovery exacerbate women's vulnerability, but other gender-blind government policies can also increase women's vulnerability by reducing their access to land and thereby their resilience to disaster. In Chap. 5 titled 'Urbanization and disaster: Loss of women's property ownership in Leh, Ladakh', *Bhuvaneswari Raman* studies the changing property rights of women amidst rampant urbanization in a high-risk ecological zone. Buddhist women in Leh town in the Indian state of Jammu and Kashmir traditionally had some land rights under specific circumstances. The form of patriarchy practiced in this Buddhist community gave women more freedom than that practiced among the patriarchal Hindu and Muslim cultures of northern India. However, such land rights among Buddhist women are threatened by government policy on tourism and urban development in the area, which makes land a highly contested resource of great value. So, women have become more vulnerable to risks than men, with their weaker access to resources—that is, land. At the same time, government tourism policies have put the area at high risk for disasters. As discussed before, women's weak land rights make women's experience of disaster disproportionately more severe than men's, and that is what is happening in Leh. The area has steep slopes with old and dilapidated buildings built along narrow lanes. It is vulnerable to landslides especially during earthquakes. The government's tourism development plan does not take into consideration disaster mitigation.

Women are not always at the losing end despite such gender-blind policy and relief programs, and they display considerable agency as *Ramanie Jayatilaka's* chapter demonstrates. In her chapter titled 'A coir mill of their own: Women's agency in post-tsunami Sri Lanka', *Jayatilaka* studies women's livelihoods in post-tsunami Sri Lanka, and how despite being ignored by relief programs, which focused on men's livelihoods, women have come together to create their own space and livelihoods. Weaving coir ropes has been a traditional source of home-based employment for women in the coastal Galle district of Sri Lanka. After the tsunami, national and international attention focused on fishermen and their lost livelihood, whereas women's livelihoods such as coir weaving, making and selling food, were ignored. Support was largely given either to land-based livelihoods or to fisheries related activities and the coir weaving activities of women were excluded. Businessmen lost interest in rebuilding the coir business from which they had earlier derived profits. Their attention shifted to fisheries which seemed more profitable given the vast amounts of aid money pouring in. A women's organization stepped in and by building solidarity with other women's organizations, they helped women to revive the coir weaving sector, and restore their livelihoods. They also set up a coir mill which was owned and operated by the women. Given the cultural barriers to women's mobility, and practical constraints such as caring for family and children, it was a challenge for women to source coconut husks, and deliver finished goods to the market. In addition, cheaper imports threatened the market and reduced the value of women's labour. Women have met these challenges and persisted and continue to run the mill to the best of their capacity. In this instance, the indifference of the state, international donors and businessmen to women's non-land based post-disaster livelihood left them poor and unable to earn an income. However, with support from women's organization, women were able to rebuild their livelihood.

The last chapter highlights the importance of context and various axes of disadvantage faced by women. *Carol Yong, Frans R. Siahaan*, and *Andreas Burghofer* titled "Gender, land tenure, and disasters in the Mentawai Islands, Indonesia", presents the complex situation in Mentawai, Indonesia. The authors identify the link between gender, natural resource, land tenure, and disaster and argue that nuances around gender and other categories are often ignored. Logging and oil palm plantations on the Mentawai islands have reduced the natural resource base of the indigenous people, and thereby their ability to mitigate and recover from disasters. However, in the context of Mentawai's patrilineal social structures and customs, women's position is further weakened by external influences of individuals, policies and corporates. Since women have little right to land, in the event of a disaster, recovery processes are gender differentiated. Development agendas and interventions construct the vulnerabilities of women and men differently. Since the Mentawai islands are governed by bureaucrats from mainland Indonesia, who tend to systematically marginalise indigenous people and privilege outsiders, the basic human rights of the Mentawaians are not protected. Thus Mentawai women were doubly disadvantaged by ethnicity and gender.

1.7 Hopes and Strategies for Empowerment

These cases show that women in different parts of Asia have different land entitlements. In some places, women had access to land, but not land titles. In other places, they had control over land. However, disaster—natural disaster or human-made disaster, or even the threat of disaster—have disturbed such land entitlements. Women have either fought for land rights and lost, as in Sarawak; or chose to live in fear instead of giving up their land, as in Nawalapitiya. In some cases, the changes and losses have been so subtle that women cannot even protest against it, as seen in Leh, Ladakh and West Sumatra, Indonesia. This is especially when the struggle is against the state's plans for the region, and the local people's rights are ignored.

The cycle of disaster has built on women's existing vulnerability and disadvantages. Not only have they been deprived of land rights, but their voicelessness and heavy household responsibilities have alienated them from most of their entitlements. Hence, in the subsequent cycle of disaster, they will be starting from a much weaker position than before, intensifying their vulnerability as the cycle of disaster continues. What is of particular importance is that gender-blind development policies and risk mitigation and relief efforts of government and non-governmental organizations can further exacerbate this vicious circle by introducing a patriarchal land regime. Lack of understanding about local practices and contexts can intensify and solidify existing gender inequities. These case studies illustrate how both NGOs and the State are further impoverishing women and increasing their vulnerabilities by denying their land entitlements. Intersecting disadvantages and deprivations of gender, class, caste, location and community then multiply the impact of the disaster.

However, the chapters also offer some hope in terms of how such a negative spiral may be mitigated and reversed, as seen in the case study on the coir industry in Sri Lanka. To understand gendered vulnerability during disaster, we need to take into consideration the intersecting web of deprivation that surrounds women, anchored in land, and other resources for livelihood. Since disasters in Asia are expected to increase in the coming decades, it is time for a new policy framework for disaster policy to ensure that disaster preparedness, mitigation and recovery and reconstruction are all informed by existing gendered practices on land entitlements and have clear focus in overcoming and reducing gendered vulnerabilities.

References

Abdulharis, Rizqi, 2006: "Land Administration in Post-Disaster Areas: Case Study of Banda Aceh, Indonesia" (MSc dissertation, Delft University of Technology).

Agarwal, Bina, 1994: *A Field of One's Own: Gender and Land Rights in South Asia* (Cambridge: Cambridge University Press).

Agarwal, Bina, 2003: "Gender and Land Rights Revisited: Exploring New Prospects via the State, Family and Market", in: *Journal of Agrarian Change*, 1,1–2 (London: Blackwell).

APWLD, 2005: "Why are Women more Vulnerable During Disasters? Violations of Women's Human Rights in the Tsunami Aftermath". Asia Pacific Forum on Women, Law and Development (APWLD), Thailand (October); at: http://iknowpolitics.org/sites/default/files/tsunami_report_oct2005.pdf.

Brown, Oli; Crawford, Alec, 2006: *Addressing Land Ownership after Natural Disasters—An Agency Survey* (Winnipeg: International Institute for Sustainable Development).

Chagutah, T., 2013: "Land Tenure Insecurity, Vulnerability to Climate Induced Disaster and Opportunities for Redress in Southern Africa", in: *Jàmbá: Journal of Disaster Risk Studies* 5,2.

CSRC (undated): Women Land Rights Campaign, Community Self Reliance Centre, Kathmandu; at: http://www.csrcnepal.org/index.php.

de Mel, Neloufer; Ruwanpura, Kanchana N.; Samarasinghe, Gameela (Eds.), 2009: *After the Waves: The Impact of the Tsunami on Women in Sri Lanka* (Colombo: Social Scientist Association).

Enarson, Elaine; Chakraborti, Dhar P.G., 2009: *Women, Gender and Disaster: Global Issues and Initiatives* (New Delhi: Sage).

Enarson, Elaine; Morrow, Betty Hearn (Eds.), 1998: *The Gendered Terrain of Disaster: Through Women's Eyes* (Westport: Praeger).

Fandl, Kevin J., 2005: "Dead Capital and the Sea: Post-Tsunami Relief for the Southeast Asian Informal Economy", in: *Yale Journal of International Affairs*, 79–88.

FAO, 2005: *Gender and Land Compendium of Countries Studies* (Rome: FAO).

FAO, 2007: *Gender and Law: Women's Rights in Agriculture* (Rome: FAO).

Fricska, Szilard; Leon, Esteban; da Passano, M. Guglielma, 2010: "Land and Natural Disasters—Guidance for Practitioners" (Nairobi: United Nations Human Settlements Programme [UN-HABITAT]).

Fitzpatrick, Daniel, 2008: *Women's Rights to Land and Housing in Tsunami-Affected Aceh, Indonesia*, ARI working paper No. 3 (Singapore: Asia Research Institute).

Ginnetti, Justin; Dagondon, Butch; Villanueva, Clarisse; Enriquez, Juanito; Temprosa, Francis Tom; Bacal, Christie; L., Fr. Norberto; Carcellar, C.M., 2013: *Disaster-Induced Internal Displacement in the Philippines: The Case of Tropical Storm Washi/Sendong* (Chatelaine: Internal Displacement Monitoring Centre).

Handmer, John; Choong, Wei, 2006: "Disaster Resilience Through Local Economic Activity in Phuket", in: *The Australian Journal of Emergency Management*, 21,4.

Hewitt, Kenneth, 1997: *Regions of Risk: A Geographical Introduction to Disaster* (Harlow: Longman).

Holmgren, Teea Emilia; Moe, Anne Katrine, 2012: *Disaster Risk Reduction in Liberia: How do International Humanitarian Organizations Involve and Promote Local Capacities in Disaster Risk Reduction in Liberia; and What is the Local Women's Role?* (University of Stavanger, Norway); at: http://www.gdnonline.org/resources/Holmgren%20T%20E%20%20%20Moe%20A%20K%20%20(1).pdf.

Jackson, Cecile, 2003: "Gender Analysis of Land: Beyond Land Rights for Women?", in: *Journal of Agrarian Change*, 3,4: 453–480.

Julia; White, Ben, 2012: "Gendered Experiences of Dispossession: Oil Palm Expansion in a Dayak Hibun Community in West Kalimantan", 89n: *The Journal of Peasant Studies*, 39,3–4: 995–1016.

Kabeer, Naila, 2009: *World Survey on the Role of Women in Development: Women's Control over Economic Resources and Access to Financial Resources, Including Micro-Finance* (New York: UN-DESA).

Kachingwe, Nancy, 2012: "From Under Their Feet: A Think Piece on the Gender Dimension of Land Grabs in Africa" (London: Actionaid).

Kandiyoti, Denis, 1988: "Bargaining with Patriarchy", in: *Gender and Society*, 2,3: 271–290.

Knight, Rachael; Adoko, Judy; Auma, Teresa; Kaba, Ali; Salomao, Aldo; Siakor, Silas; Tankar, Issufo, 2012: *Protecting Community Lands and Resources. Evidence from Liberia, Mozambique and Uganda* (Rome: IDLO); at: http://issuu.com/idlonews/docs/communitylandreport.

Kusakabe, Kyoko; Vongphakdy, Sengkham, 2014: "Gender Vulnerabilities of Resettlement and Restricted Mobility of Ethnic Groups in Northern Laos", in: Lund, Ragnhild; Kusakabe, Kyoko; Mishra Panda, Smita; Wang, Yunxian (Eds.): *Gender Mobilities and Livelihood Transformations: Comparing Indigenous People in China, India and Laos* (London: Routledge): 134–165.

Land Watch Asia, 2012: "Climate Change can Deal Blow to Land Tenure" (Manila: Asian NGO Coalition for Agrarian Reform and Rural Development); at: http://landwatchasia.wordpress.com/2012/08/09/climate-change-can-deal-blow-to-land-tenure/ (30 November 2013).

McClean, Denis (Ed.), 2010: *World Disasters Report 2010: Focus on Urban Risk* (Geneva: International Federation of Red Cross and Red Crescent Societies).

Mehta, Manjari, 2007: *Gender Matters: Lessons for Disaster Risk Reduction in South Asia* (Kathmandu: ICIMOD).

Meinzen-Dick, Ruth; Brown, Lynn; Feldstein, Hilary; Quisumbing, Agnes, 1997: "Gender, Property Rights, and Natural Resources", in: *World Development*, 25,8: 1303–1315.

Mitchell, David, 2010: "Land Tenure and Disaster Risk Management", in: *Land Tenure Journal*, 1: 121–141; at: http://www.fao.org/nr/tenure/land-tenure-journal/index.php/LTJ/article/view/11/5.

Pincha, Chaman, 2008: *Indian Ocean Tsunami Through the Gender Lens: Insights from Tamil Nadu, India* (Mumbai: Earthworm Books).

Rao, Nitya, 2006: "Land Rights, Gender Equality and Household Food Security: Exploring the Conceptual Links in the Case of India", in: *Food Policy*, 31: 180–193; at: http://www.sciencedirect.com/science/article/pii/S0306919205000862.

Rao, Nitya, 2011: "Women's Access to Land: An Asian Perspective", Paper presented at the expert group meeting 'Enabling rural women's economic empowerment: institutions, opportunities and participation' at Accra, Ghana, 20–23 September 2011.

Razavi, S., 2007: "Liberalization and the Debates on Women's Access to Land", in: *Third World Quarterly*, 28,8 (London: Routledge).

Rees, Susan; Pittaway, Eileen; Bartolomei, Linda, 2005: "Waves of Violence—Women in Post-Tsunami Sri Lanka", in: *Australasian Journal of Disaster and Trauma Studies*, 2005,2; at: http://www.massey.ac.nz/~trauma/issues/2005-2/rees.htm.

Ruwanpura, Kanchana N., 2008: "Temporality of Disasters: The Politics of Women's Livelihoods "After" the 2004 Tsunami in Sri Lanka", in: *Singapore Journal of Tropical Geography*, 29,3: 325–340.

Scalise, Elisa, 2009: *Women's Inheritance Rights to Land and Property in South Asia: A Study of Afghanistan, Bangladesh, India, Nepal, Pakistan, and Sri Lanka* (Seattle: RDI).

Schlager, E.; Ostrom, Elionore, 1992: "Property Rights Regimes and Natural Resources: A Conceptual Analysis", in: *Land Economics*, 68,3: 249–262.

Shaw, Carole; van Unen, Judith; Lang, Virginia, 2012: *Women's Voices from the Flood Plains: An Economic Gender Lens on Responses in Disaster Affected Areas in Queensland and Victoria*. Economic Security 4 Women and Justice Equality Rights, Access International, Australia; at: http://www.security4women.org.au/wp-content/uploads/eS4W_Womens-Voices-from-the-Flood-Plains_Report.pdf.

Trohanis, Zoe Elena; Svetlosakova, Zuzana, Carlsson-Rex, Helene, 2011: *Making Women's Voices Count in Natural Disaster Programs in East Asia and the Pacific*, EAP DRM Knowledge Notes Series No. 24 (Washington, DC: World Bank); at: https://openknowledge.worldbank.org/handle/10986/10091.

Usamah, Muhibuddin, 2013: "Land Tenure Security and Resilience to Multiple Disasters: A Study of Camalig Municipality, Province of Albay, The Philippines" (Ph.D. dissertation, Royal Melbourne Institute of Technology (RMIT) University, Australia).

Villarreal, Marcela, 2006: "Changing Customary Land Rights and Gender Relations in the Context of HIV/AIDS in Africa", Colloque international "Les frontières de la question foncière—At the frontier of land issues", Montpellier; at: https://www.mpl.ird.fr/colloque_foncier/Communications/PDF/Villarreal.pdf.

White, Ben; Borras Jr., Saturnino M.; Hall, Ruth; Scoones, Ian; Wolford, Wendy, 2012: "The New Enclosures: Critical Perspectives on Corporate Land Deals", in: *Journal of Peasant Studies*, 39,3–4: 619–647.

Wisborg, Poul, 2014: "Transnational Land Deals and Gender Equality: Utilitarian and Human Rights Approaches", in: *Feminist Economics*, 20,1: 24–51.

Wisner, Ben; Blaikie, Piers; Cannon, Terry; Davis, Ian, 2004: *At Risk: Natural Hazards, People's Vulnerability and Disasters* (London: Routledge, 2nd Ed.).

Chapter 2
Gender Impact of Large-Scale Deforestation and Oil Palm Plantations Among Indigenous Groups in Sarawak, Malaysia

Carol Yong and Wee Aik Pang

Abstract Land and forest ecosystems form the core of the belief systems and daily lives of indigenous forest people and communities. However, State policies and laws introduced in the colonial period, retained and reinforced by post-colonial states have substantially increased the state's power and are restricting and removing indigenous rights to land and forest resources according to *adat* (traditional customs). This chapter examines the impact of changing land use and land tenure systems in Sarawak on human rights, livelihoods, and local gender practices. Conversion of forests to oil palm plantations is regarded as a disaster given the importance of land for customary practices, food security and income-generating activities, and other fundamental rights of indigenous peoples. We use the term 'disaster' from a variety of perspectives, foremost is the communities' perspective placed alongside other perspectives such as gender, legal, socio-cultural, economical, and environmental. To support these arguments, this chapter studies the Iban community of Kampong Lebor whose customary lands were cleared by companies to plant oil palm without free, prior, and informed consent (FPIC). Large-scale plantations on these lands contributed to significant social and environmental risks and other negative socio-economic and climatic consequences. A human-made disaster in Sarawak was partly averted by restoring traditional land rights and tenure systems; however, without restoring women's access to forest.

Keywords Sarawak · Oil palm plantations · Land grabbing · Customary rights · Disaster

Abbreviations

BN	Barisan Nasional
FPIC	Free, prior, and informed consent
Kpg.	Kampong

Carol Yong, independent researcher, Malaysia, Email: yrn@vjfg.at; Wee Aik Pang, independent researcher, Malaysia. Email: weeaikpang@gmail.com.

K. Kusakabe et al. (eds.), *Gender and Land Tenure in the Context of Disaster in Asia*, SpringerBriefs in Environment, Security, Development and Peace, DOI 10.1007/978-3-319-16616-2_2

LCDA	Land Custody and Development Authority, Sarawak
NASA	National Aeronautics and Space Administration
NCR	Native customary rights
NGO	Non-governmental organization
PBB	Parti Besaka Bumiputra
SACCESS	Sarawakians Access
SOPPOA	Sarawak Oil Palm Plantation Owners Association
UN	United Nations

2.1 Introduction[1]

Current studies using satellite-based imaging, NASA data and other tools reveal that only small areas of intact forests remain in Malaysia, especially Borneo, because much of the forests have been heavily logged or cleared for timber or oil palm production (Bryan et al. 2013; Hansen et al. 2013). Experiences of many Sarawak landowners reveal that land acquisition by the state authorities to feed hungry plantation owners, loggers and developers is threatening the survival and livelihoods of indigenous peoples and local communities. Many of these lands are held under customary tenure by local communities, and over 200 cases have been challenged by NCR landowners in courts. Too often, the State cites poverty eradication, development and modernization of agriculture to justify land acquisition and oil palm plantation development, specifically targeting *native customary rights* (NCR) lands, and lately, using dam building as the only way to develop remote areas.

To identify the link between oil palm agribusiness land-grabs and disasters, our starting point is that land and forests form the core of the belief systems and daily lives of indigenous forest people and communities in Sarawak and other parts of Malaysia as well as globally where indigenous peoples live. From this perspective, customary rights and relationships with their traditional land uses have existed long before statutory regulations. We also know, as told by communities, that lands are continuation of generations and their lands are filled with stories, good and bad, memories, boundaries, sacred places, mountains that gave birth to legends, and so on. But what is happening on-the-ground through communities' experiences, and more generally research/fieldwork/documentation by NGOs, researchers etc., we see communities' forests and customary lands continuously exploited through

[1]We are grateful for the comments, insights, criticism, and support provided by SACCESS and land rights lawyer and Sarawak State MP, See Chee How. We are also grateful to the anonymous reviewers who commented on earlier versions of this case study.

concessions and licenses issued for development and traditional land uses being completely disregarded and ignored by the so-called modern state, throwing into complete doubt even the whole definition of indigenous communities. Here we are talking about communities being uprooted, destroyed and much of what is lost can never be recovered. Thus in fact saying development-induced land and other problems for communities is a disaster is putting it mildly.

Secondly, the UN and other international and domestic laws and human rights treaties have validated the rights of indigenous peoples to self-determination including their rights to land. Yet, under current power relationships, it also becomes obvious that the state is in cohort with capitalists, whilst in the Sarawak/ Malaysia context such relationships are reportedly connected to practices of cronyism and nepotism (cf. http://www.sarawakreport.org; Ross 2001; Yong 2010). With the Federal court reaffirming landmark cases granting NCR rights, the human-made disaster brings to forth how the executive is refusing to acknowledge the third pillar of democracy, the judiciary, thus seriously undermining the whole notion of democratic institutions and practices.

Economically, the once self-sufficient, subsistence communities have since turned into wage earners to sustain themselves. Without lands to cultivate and grow staple food such as rice, hunt and collect jungle produce, communities are increasingly dependent on the cash economy and jobs available, but opportunities are limited for rural families. What we then see on the ground is a situation where rural folks are becoming more dependent on outsiders and outside system—a survival disaster perhaps, as once independent communities continue to be forced into dependency.

Likewise, environmentally, degraded lands, polluted rivers and streams, and land clearing methods all combine to alter the natural environments and ecosystems. Once clean rivers and streams meeting the water needs of communities have since become unsuitable for human use, bringing diseases when people have little or no choice but to use polluted water. During the dry season, there is no rain to collect water to supplement needs. Mostly women bear the burden of meeting the family's water needs.

This chapter first looks briefly at the context of the site of the study, namely Sarawak and an Iban village called Kampong (Kpg.) Lebor. It then clarifies the methodology and the terms used referring to customary land tenure and traditional land uses based on communities perspectives. Next, the chapter analyses changes affecting customary lands, in particular Sarawak's large-scale oil palm plantations which we consider a disaster for indigenous peoples and for women. Here we present the case of Kpg. Lebor to illustrate how the community is addressing the issue of non-FPIC appropriation of their customary land and reclaiming their rights. The chapter ends with a holistic and practical understanding of land as held by the Kpg. Lebor people as opposed to the state's view which focuses on land as an economic asset to be exploited for profit generation. This changing view has a huge impact on women who are primarily responsible for the well-being of the family and community.

2.2 The Context: Sarawak and Kampong Lebor

Sarawak is Malaysia's largest state with a total land area of 124,450 km^2. It was ruled by the Brooke family for 100 years (1841–1941).[2] After a brief period of Japanese occupation (1941–1945), Sarawak and Sabah became British colonies (1946–1963). In 1963, Sarawak together with Sabah and Singapore formed Malaysia with 11 other states. (Singapore became an independent republic in 1965.)

In 2010, Sarawak's total population was about 2.47 million (Census 2010). Its population density is low at 22 persons per sq. km. The 28 indigenous groups and subgroups in Sarawak are collectively called Dayak and Orang Ulu. They compose over 50 % of the population in Sarawak. The Iban are the largest indigenous group (30.3 %) followed by the Bidayuh (8.3 %). Other groups include Kenyah, Kayan, Penan, Melanau, Bhuket, Lun Bawang, and Punan Bah. Chinese, Brunei-Malays, Malays, Indians, and Eurasians also form part of the population.

Most indigenous people live in the rural and interior regions and practice a range of traditional livelihoods that depend on land and forests and their ecosystems. These include fishing, hunting, and gathering or collecting forests products such as timber, rattan, fodder, jungle fruits and vegetables, and other materials essential for their existence. Food is derived from *padi* (rice) cultivation, mixed vegetable gardening, and rearing chicken. Income is derived from smallholding rubber plantations or paid employment. Some women also sell handicrafts for cash.

Kampong Lebor (henceforth Kpg. Lebor) is an Iban settlement of 186 people.[3] It is located in the district of Serian in Samarahan division, about 70 kms from Kuching, the capital of Sarawak. The Iban community of this village is the eighth generation now occupying and cultivating the *native customary right* (NCR) lands.

Information was obtained through direct involvement and observations, community dialogues, interviews with community leaders, organising communities, focus-group and individual discussions/interviews. Information was also collected from court documents, oral history, court proceedings, etc. by a researcher, who has worked with the communities for a long time. Our personal and professional experiences and understanding of these themes and issues is supported by relevant literature.

---

[2]The Brooke family, starting with James Brooke, installed themselves as the White Rajahs (rulers) of Sarawak. James Brooke, a middle-class English adventurer, was rewarded with a block of territory (now the First Division of Sarawak) by the Sultan of Brunei in 1840 for helping to suppress a miners' rebellion against the Sultanate. He acquired further territories and by 1864, the entire area under the control of the Brunei Sultanate was recognized as the independent state of Sarawak. James Brooke was succeeded by his nephew Charles Brooke as Second Rajah (1868–1917). Charles Vyner Brooke became the third Rajah in 1917 until the end of the Brooke's rule in 1941.

[3]Information on Kpg. Lebor is used with the permission of the villagers, supplemented with information obtained from the NGO SACCESS and See Chee How, a Sarawak state MP and land rights lawyer.

The Kpg. Lebor community and the Sarawak NGO SACCESS have forged a partnership since 1998 when the case was first filed. It has been a lengthy, costly and dangerous struggle to defend their rights to NCR land, including arrest by police, threats by gangsters, and punishment by the government. Even today, people opposing plantations face police investigations after reports are lodged against them by a minority camp within the community, accusing them of siphoning off money, etc.

After winning the court case, the community wants to learn how to develop their lands and regain full control. So Kpg. Lebor is more than just a successful case to show community resistance to agribusiness expansion. It is a community that wants to develop themselves, while remaining deeply rooted in their social, cultural, traditional, economic and political roots. Ultimately, this means the community wants to regain their independence and self-determination in social, economic and political positions as before, but with modern agricultural business and other sustainability options that are being explored with experts. Furthermore, such determination from the community demonstrates a stark contrast to the state model of development in which people are placed as passive and unknowledgeable 'natives' that must have 'saviours' from the state and capitalists private sectors to develop their idle unproductive NCR lands, a biased and non-objective claim at best (see, Matsubara 2003). For these reasons, Kpg. Lebor makes an interesting case study, whereby multiple interests, frictions and pressures emerge to unite or divide the community in different ways.

2.3 Customary Land Tenure

The lands of the Iban, like other indigenous groups in Sarawak, are held under what is termed customary land tenure. Based on Kpg. Lebor's genealogies and traditions, their rights over the current village area were established through first clearance and settlement by their ancestors a long time ago. Their ancestors, of Remun Iban origins, were one of the earliest Iban migrants from the Kapuas Basin,[4] who moved into the area, felled sections of unexploited primary forests, and cultivated the land. Thus, they acquired NCRs to that particular territory, referred to in Iban as *pemakai menoa*, which also encompassed the higher forests (*pulau galau*).[5]

---

[4]In pre-colonial times, an area of the Island of Borneo was known as Kapuas Basin. European colonial powers partitioned Borneo into Kalimantan under Dutch rule and Sarawak (initially under the Brooke family and subsequently) as a British crown colony, along with Sabah then called North Borneo.

[5]While each indigenous group has its own concept and terms, this chapter uses the Iban's explanation of *native customary right* lands (NCR lands) or *native customary lands* (NCL). The Iban concept of NCR was affirmed in the High Court judgement on the Kpg. Lebor case delivered in March 2012, reaffirmed in the decision of the Court of Appeal in September 2013.

Traditional territories are communally owned areas within defined boundaries that are well-known and agreed upon by neighbouring communities, normally marked by small or big valleys, small hills or ridges, rivers, streams, watersheds, or an area planted with bamboo, specific tree species, and big trees. Each village territory includes rivers, water catchment areas, and cemeteries or ancestral burial grounds which are sacred lands (*pulau mali*). Individuals and families can claim usufruct rights to farmlands (*temuda*) by clearing forests or opening up land but all villagers must first agree unanimously on this. Female and male children can inherit their parent's share of *temuda* lands, but these rights usually cease if they move out of the longhouse or break their links with the community. When this happens, the *temuda* lands revert to the community and can be granted to other community members.

For the Kpg. Lebor people, at the very least, land enables them to cultivate rice, their staple food. Rice is also used for other purposes, such as making the traditional Iban rice wine (*tuak*) used in festivals and rituals. If they did not cultivate rice, they would have to buy it but they do not have much cash in the first place as theirs is not a cash economy. The Iban recognize the importance of using *pemakai menoa* and *pulau galau* in a sustainable and efficient manner. Each household also has usufruct rights to *temuda* lands. They maintain their customs and the associated rules, inter alia:

- recognition of local land-tenure systems whereby those who first cleared lands have rights to them;
- land-use practices that rejuvenate, not harm, nature, e.g., new farming sites with sufficient fallow periods;
- strict observances for sacred sites such as ancestral graves, water sources;
- intercropping food crops; and
- selected harvesting of fruit trees and preserving specific tree species.

The Iban have been occupying and cultivating NCR lands for subsistence and trading forest products long before Sarawak was ruled by external powers. Elderly women and men inherited skills and extensive knowledge about forest resources and local oral traditions and rituals closely connected to ancestral land that they pass on to younger members through oral and practical teachings. The women collect and process rattan and make craft items out of it. Given that rattan is limited, women find practical uses for other resources like *bemban*, a local name for a type of reed. Some 30 women from Lebor are involved in craft making. Rattan and *bemban* are strongly guarded by the villagers in communal forests that are still intact. However, other parts of the forest have already been destroyed by commercial oil palm companies.

Such locally evolved practices formed part of the day-to-day life of the community and helped it conserve land and forest resources and mitigate the effects of deforestation and forest degradation. Respect for nature and security of land tenure forms the basis of their existence and livelihood. Traditional use and management of land and forests help to mitigate the consequences of climate change, global warming, extinction of plant and animal species, soil erosion, catastrophes linked to deforestation (e.g., flash floods, mudslides), and deteriorated watersheds and water quality.

Differing views on land and land tenure systems reflect two sets of priority for land uses. Indigenous people view lands and forests as sources of livelihoods, socio-cultural identity, and links between past, present, and future generations. Political and economic developers, however, prioritize commercial interests. The Kpg. Lebor community has struggled to have its land rights reinstated, demonstrating its desire to manage its lands and forests with innovative options that are sustainable, gender sensitive, and have minimal adverse effects on humans and resources.

2.4 Traditional Methods of Land Use in Kpg. Lebor

The Kpg. Lebor villagers still rely on land as a main source of livelihood, with the traditional practice of swiddening cultivation of hill *padi*, which is discussed in detail below. Some also plant crops like tobacco and rubber on a small scale to earn cash income. Meat is bought or derived from farmed livestock including a species of pig called *babi kampong*, chickens, ducks, and goats.

In hilly and low-lying areas, villagers grow rice as their staple food. For the Iban, rice cultivation is more than an economic activity; it is an elaborate system of rituals conducted at various stages of the farming cycle, starting with the clearing of the farming site and ending with the storing of the rice after the harvest. Underlying this traditional method of farming is the possession of ritual knowledge and skills, for example, on appeasing spirits to ensure bountiful harvests. Women play an important role in farming activities and rituals; they reap sacred rice, known as *padi pun*, before harvesting commences. Elderly women perform many of the elaborate rituals associated with *padi pun*, blessing the rice planted to ensure a bountiful harvest. *Padi pun* is never sold or given away, and generations of women have inherited the *padi pun*. While the senior-most man of the family presides over the offering (*piring*) ceremony and chants prayers summoning spirits to return when the rice is ready for harvesting, women pick the first strains of *padi pun* and bring them home for safekeeping until the next planting cycle. The Iban believe that the fertility of the new *padi* planted rests on the *padi pun*; thus, the first seeds are sown in the centre of the cleared farm during the new cycle.

The soil on the slopes of hilly and low-lying areas is generally poor. The traditional practice is to maintain a fallow period for regeneration of soil and forest cover. Under customary rights arrangements, traditional rights to the forest areas left fallow are vested with the family and its heirs. So, fallow forest areas are not abandoned as incorrectly perceived by the state government, but are owned by individuals or families for subsistence cultivation. These areas are also not non-productive, as the Sarawak government continues to insist and use as an excuse to bring in large-scale monocrop plantations. This policy has reduced the land available to the villagers for farming. The 'adopted' practice of chemical fertilizer and pesticides is also gaining ground and increasing costs even with government subsidy.

Often farms are far from the village, which makes daily travel difficult. Temporary huts known as *dampa* are then built close to the farms and groups of families live there for an extended period of time. Thus, rice farming is closely associated with community kinship and working together using the labour-exchange system (*bedurok*). In swidden rice farming, and in most non-rice farming activities, the entire family puts in labour. There is, however, gender variation in some of the tasks. For instance, men usually choose the site of the farms. After a site is identified, women and children clear the undergrowth and small trees while men clear the dense bush and bigger trees. Thereafter, men dibble while women sow seeds.

The traditional system of land tenure and land-use practices, as all social, economic, religious, and political aspects of Iban life, are governed by *adat* or traditional customs of the Dayak and Orang Ulu indigenous communities of Sarawak.[6] Generally, *adat* ensures a harmonious relationship among community members and maintains a general state of wellbeing in the spirit world. Breaching this *adat* would risk a breakdown in social relationships which is punishable in both the secular and spiritual senses (SACCESS 2008). Yet, the *adat* has evolved over time with communities adapting to changes around them or colonial and post-colonial governments changing it to exert control over indigenous communities and their lands and resources. Alteration of the *adat*, especially aspects of customary law on lands, or the Sarawak government's assertion that the *adat* on NCR lands does not include *pemakai menoa* and *pulau galau* inevitably affected land-use and traditional ways of life.

2.5 Changes Affecting Customary Land Tenure and Land Use

In pre-colonial times, the indigenous people of Sarawak traded many jungle products (e.g., *getah jelutong*, a milky latex that was an important ingredient of chewing gum; *damar* resins; incensed wood *gaharu*; etc.) with local and foreign traders for Chinese bronze wares, ceramics, glass beads, textiles, and large jars. These form part of family inheritances today. The forest was also a resource base for the family's basic needs: food, fuel, water, and raw materials such as rattan for their traditional crafts.

The natives' NCR was first legally 'codified' under James Brooke, the first White Rajah, who seemingly understood that land was central to the way of life of Sarawak's indigenous people practicing swidden cultivation.[7] Thus, in the 1812 Code of Laws, immigrants were only allowed to settle on land not already occupied by natives. NCR was recognized under the following conditions:

---

[6]Colchester et al. (2007: 9) defined 'custom' (*adat*) as a body of beliefs, social norms, customary laws and traditional practices passed on from one generation to the next as oral tradition.

[7]That meant first converting non-written into a written law with governmental statutory power. While previously localized *Adats* held power over lands, usage and other aspects tied to life, not as how statutory laws were compartmentalized with one law each for lands, forests, agriculture, and so on up to now. Effectively, colonial law did not recognize NCR but rather destroyed *Adats*.

- the felling of primary forests and the occupation of the land thereby cleared;
- the occupation or cultivation of land;
- the use of land for burial grounds or shrines; and
- the use of land of any class for right-of-way.

Nevertheless, the past century saw gradual changes to land tenure and land use, especially forests. Land policies and legislations were instituted by British colonizers to exploit resources to boost the colonial economy. Two key colonial legislations on forests and lands were the Forest Ordinance 1953 and the Sarawak Land Code 1958. The Forest Ordinance brought ownership and use of forests in Sarawak under the jurisdiction of the state forest department. The Sarawak Land Code enacted certain provisions that curtailed any NCR land claims after 1958. NCR land was, as it is today, defined as land in which NCR, whether communal or otherwise have lawfully been created prior to January 1, 1958. Changes in land use also occurred with the shift from subsistence farming to cultivating cash crops. Further changes in land use resulted from the development of infrastructure such as roads (for marketing and transportation of cash crops) and new settlements, facilitating access for migrant populations.

The post-colonial state government followed much of the same policies and legislation. The Sarawak Land Code (Cap 81), the main land legislation, was amended frequently to limit NCR lands (SACCESS 2008). Amendments in 1994 empowered the Land Minister to abolish NCR to lands. In 1996, the burden of proof for NCR over land was put on native landowners who had to prove that they had been on the land before January 1, 1958, and had exercised their *adat* to clear forests, cultivate, live, and occupy the lands and continued to do so. The 2000 amendment removed the category ‘any other lawful method’ from Article 5(2), allowing land-grabbing of native lands that were then handed over to private developers. These amendments contradicted the land code and the specific stipulations recognizing NCR in written law under Section 2(a) and through the classification of lands into Mixed Zone, Reserved Land, Native Area Land, Interior Area Land, and Native Customary Land.

2.6 Sarawak State Government Priorities

Private sector involvement in oil palm plantations was encouraged through the ‘New Concept’ or *Konsep Baru* land development policy of 1996 engineered by the Sarawak government under Taib Mahmud.[8] Under the scheme, native landowners were expected to surrender their lands to the State for 60 years. This was to be held in trust for native landowners by an appointed state agency. The state agency and private companies then formed joint-ventures for oil palm schemes. In theory, native

[8]As of March 1, 2014, Taib had been appointed the new governor of Sarawak, a day after he resigned as the Chief Minister after 33 years in office. Taib appointed his former brother-in-law, Adenan Satem, to succeed him as Sarawak’s fifth Chief Minister.

landowners can apply to reclaim their lands on expiry of the lease. In reality, the land rights of communities are now uncertain; recovery of their land is not guaranteed.

Holding three key posts (Chief Minister, Finance Minister, and Planning and Resources Management Minister) for a long time gave Taib excessive control over land, timber, minerals, and other resources. He was directly responsible for handing out leases, concessions, and licenses for commercial plantations, logging, and privatization of infrastructure projects such as dams and roads (IDEAL 1999; Yong 2010). Investigations by several international NGOs have exposed his involvement in corruption, tax evasion, and land grabbing through his control over licenses on logging and oil palm plantations in Sarawak.[9] FPIC as a precondition for exploration, development, exploitation, and utilization of natural resources in areas occupied by indigenous peoples was ignored. Large-scale industrial logging and clearing of forests for commercial oil palm plantations in Sarawak resulted in conflicts with the indigenous people that still continue. Hundreds of landowners have filed cases in courts over the past three decades under Taib's rule (see below, the Kpg. Lebor case).

Sarawak's Land Development Minister, James Masing, claims that the government's intent is to double the area of land under oil palms, currently around 1 million hectares, to 2 million hectares by 2015, while the Sarawak Oil Palm Plantation Owners Association (SOPPOA) accidentally disclosed that the government's actual goal is 3 million hectares.[10] Increasing global market demand, consumption, and trade in bio-fuels have led countries like Malaysia to respond eagerly with forest conversion, especially in Sarawak and Sabah, for commercial oil palm plantations to facilitate related exports.

2.7 Large-Scale Oil Palm Plantations—Environmental Risks and Other Impacts

Around 1996–1997, the Lebor villagers found out that the Sarawak government had chosen the Sarawak *Land Custody and Development Authority* (LCDA) and a private company to start clearing lands and planting oil palm seedlings on their customary lands. Bulldozers were used to clear the area, destroying the villagers' sources of food, fodder, water, valuable medicinal plants, wildlife, and forest products. Damage was also inflicted on fruit trees, rubber groves, and other essential trees and crops that provided the villagers food and income. In other words, local economies were destroyed. This added the burden onto women to find alternatives since women traditionally assume these roles.

---

[9]For details about Taib and the companies owned by him or members of his family, numbering over 400 companies in 25 countries and offshore jurisdictions, see at: http://hornbillunleashed.wordpress.com/, http://www.sarawakreport.org, http://www.bmf.ch. The family's stake in 14 Malaysian companies alone is over USD1.46 billion (MYR4.6 billion). See http://www.bmf.ch, http://www.stop-timber-corruption.org for additional information.

[10]*Sarawak Report*, January 20, 2014.

Clear-felling of vast tracts of forest land also resulted in biodiversity loss, soil impoverishment, and depletion and pollution of natural habitats and water resources. The river was polluted by soil erosion and waste from land-clearing activities, which, in turn, affected the villagers' water source, fish, and other riverine resources. These hazards posed serious health problems for the villagers, again increasing women's workload as the primary caregivers.

Monoculture species are more susceptible to pests and diseases than indigenous species. However, the use of agrochemicals to control pests and diseases could contribute to further long-term risks to humans and ecosystems. Experiences of other communities forced off their lands and employed as plantation labourers revealed the heavy or widespread use of chemical pesticides and fungicides to control pests and diseases, common with mono-crops such as oil palm. Women and men from poorer families are involved in weeding. They also tend to suffer more health-related effects because of the lack of resources for nutrition as well as medical and other needs. The companies did not seem to be bothered about these negative effects on the villagers.

Led by their then headman, TR Nyutan, the community made several attempts to meet with government and company officials and lodged nine police reports to oppose the encroachment. The authorities did nothing except to tell them the government had issued permits to the LCDA and the company to develop the disputed lands. In effect, the government completely disregarded the land rights of the people of Kpg. Lebor and issued land leases to the company.[11] The villagers were also denied compensation. The only option left was to access legal means to defend their rights.

2.8 Kpg. Lebor: A Success Story

On November 24, 1998, on behalf of his community, TR Nyutan and few other villagers as plaintiffs filed a case in the Kuching High Court.[12] However, after the case was filed in the Kuching High Courts, a few villagers began distancing from the main group that continued to protest against land-clearing works. Throughout

[11]Research by SACCESS found that the private company Nirwana Muhibbah Sdn Bhd was wholly owned by a ruling *Barisan Nasional* (BN) State Legislator, Mohd. Naroden B. Haji Majais, and his family. He was also an Assistant Minister with multiple portfolios in the Sarawak Government, in Planning and Resource Management, Assistant Minister in the Chief Minister's Department, Assistant Minister of Land Development, Sarawak, and Assistant Minister for Entrepreneurial Development. He is a central committee member of *Parti Besaka Bumiputra* (PBB), the main component of the ruling BN alliance. It is alleged that since Naroden is a very powerful and influential member in both the Sarawak Government and the ruling BN, no action was taken against him by the authorities or the police.

[12]High Court Suit No. 22-249-98-III (I) Tuai Rumah Nyutan ak Jami and two others vs the LCDA and two others. From this case study, it is argued that the State as in Sarawak Government refused to acknowledge land rights and lease out lands to private entities and communities have to sought rights through the courts.

the main group's long struggle, both in court and on the ground, the minority had stayed away. The minority group started trying to assert its authority, with the headman's (appointed in 2005 after TR Nyutan resigned, see below) involvement, after the High Court victory and when communities started benefitting financially from the sale of palm fruits. Interpreted in the broader context, this is another typical and often neglected dimension of both land-grabs and disasters. Companies seek to create divisions among community people with payments etc., to generate or build up the sense of creating potential 'winners' as well as losers, and, as such increasing companies' bargaining power, while also creating rifts in communities.

Land clearing for planting expanded closer to village land; even land with timber was cleared. After extracting valuable timber from the villagers' NCR lands, the company planted oil palm in the area. Villagers alleged that gangsters were employed to intimidate the villagers, particularly their leaders and strong supporters. The community retaliated by coming together for meetings, attending trials, and putting up blockades. A few villagers were against filing a court case and supported the government and the company. This issue is not within the scope of this chapter, so we restrict ourselves to pointing out that companies often give landowners verbal assurances of providing jobs once the plantations are opened. This is to ensure their cooperation. However, later, they realize that they are only hired as contract workers with daily fixed-rate wages; and often they are replaced by cheaper Indonesian labor.

With no resolution in sight and the plantation edging nearer to the village, the women and men asserted their rights by removing oil palm seedlings planted by the company on their NCR lands. Further police reports were lodged against the company but no action was taken. In contrast, when the company reported that the villagers were removing their oil palm seedlings, the police immediately responded and began an official investigation. Constant pressure to back the 'development' forced TR Nyutan to resign and the villagers' situation worsened in 2005 with the appointment of a new village head, backed by a local incumbent politician involved in the land dispute.

In the meantime, there were internal conflicts in the community as well. Once the people discovered that their customary land was taken over by oil palm plantations, they approached the late Kuching MP, Sim Kwang Yang. Then NGOs and concerned individuals (e.g. See Chee How, now a prominent land rights lawyer and state MP) followed up the community's call for support to defend and reclaim their land rights and then the court case with lawyers. Then, oil palm companies also started developing strategies to weaken opposition to the plantations. Villagers allege the company's involvement in creating internal village conflict, for example, paying and promising certain benefits to some people. As a result, despite the strong opposition to the occupation of land by the oil palm plantation, a minority of the indigenous people support the plantation company. However, even those who oppose the struggle for land rights want to claim the benefits that come from reclaiming plantation land. Such men would ask their wives to collect the money being disbursed. Such utilization of women to gain benefit suggests the invisibility and lower status of women in public space. Women's actions are not taken as seriously as that of men, and hence women take the money from community harvesting, while the men maintain the public stance of opposing the majority of villagers who are managing the returned portion of NCR lands. Basically, a minority of villagers are opposed to the majority managing and distributing benefits.

Although men were more visible in this struggle, directly confronting the government and the company, women were equally active in the struggle, attending court hearings, facing the police and company officials, and participating in blockades. Women also took care of the children and elderly women and men in the community when men attended court hearings. Different gender roles were adopted because the emphasis was on the need for the community to organize and strengthen itself against land grabbers. Land tenure was seen as a serious community issue and strategies adapted to local needs and situations.

The trial of this case ended in 2006. After a long wait, the Kuching High Court, in March 2012, delivered a judgment declaring the Lebor villagers rightful owners of about 4000 ha of lands. Disputed plantable lands, meaning those NCR lands included in the provisional lease issued by the state to oil palm companies and cultivated with oil palm (excluding roads, storage, nursery, workers' quarters and the likes), are about 1400 ha. The community had regained 300 ha of planted land which they engaged their own private company to harvest and reap the benefits from, after the court verdict in their favor. The company refused to accept the court decision and filed an appeal. Meanwhile, in August 2012, the villagers harvested around 261 acres of matured oil palm from the lands they won back after the court case. Legal recognition of the villagers' rights to the land was reaffirmed in the decision of the Court of Appeal in September 2013. This implies that the courts have held that NCR lands extend to an area wider than the *temuda*, to encompass the *pemakai menoa* (territorial domain) and *pulau galau* (reserved forests), under the Iban *adat*. However, numerous communities' experiences have revealed that instead of respecting court rulings which favor native landowners, the Sarawak state government continues to issue provisional leases and timber licences over such lands or to appeal against such judgments (Wee 2012). Thus, it is still unclear if the final decision by the courts in fully recognizing the NCR lands of Lebor villagers will be challenged in future; a final appeal has been made to the Federal Court by the company and the government.[13]

2.9 Gender Dimensions and Community Dynamics

The majority of the villages continue to oppose the land grab. Women and men, young and old are all involved in various aspects of the struggles, directly confronting the government and the company, attending court hearings, facing the

[13]In mid-2014, the judicial process was awaiting the final appeal by the companies and the State government to the Federal Court after the Court of Appeals had agreed with the High Court's decision recognizing land rights.

police and company officials, and participating in blockades, just to mention some examples. Another example is the "Action Committee" which includes both women and youths. The young equally opposed and are increasingly active in doing more paper and computer-related work as in book-keeping and accounts. Likewise, a group of young people are documenting the history of the village struggles, oral history and *adat* aimed towards a village publication.

Another issue, noted above, is that the Lebor community had access to matured oil palm from the lands they won back after the court case. This meant that the community had income from the harvests. As the court decided that the land belonged to the community, the community had since August 2012 earned monthly income by contracting out a portion of plantation to get a share of the proceeds from the 300 ha of lands returned after the High Court judgement.[14] Once the 'Action Committee' started distributing the money, villagers who were formerly opposed to indigenous struggle for land rights also wanted a share of the income. Such conflicts and deterioration in community cohesion impacted heavily on women who depend on other women in the community to complete their reproductive tasks. Moreover, the benefit of the contracted lands is paid out to the 'head of the household', usually a man. This has affected the egalitarian gender relations within Lebor households.[15]

2.10 The State, the Market and the Customary Use of Land

We have described above the politics and development policy of the Sarawak government, which has intensified acquisition of customary lands and effected unprecedented changes in customary tenure. Privatization has significantly affected security of native land tenure. It has also reduced access to and use of forest resources and lands, and as the Kpg. Lebor case revealed, removed or restricted a community's rights to its customary lands. The control exercised by the recently retired Chief Minister and current Governor (Taib) over lands and forest laws and policies has effectively negated the rights of native people with respect to customary lands and weakened their *adat*-based land tenure systems. The Kpg. Lebor community took its battle to court, where its stand was vindicated.

In Sarawak, where land-grabbing is justified by the State, only the judiciary seems capable of safeguarding these rights. The court victory for Kpg. Lebor, until

[14]Only about 300 ha of the original 1,395 ha disputed NCR lands was returned to the community. The company refused to hand over the remaining lands, pending the final outcome of the company's appeal to the highest Federal Court. The company did not pay anything at all to the community; and only returned a small part of their lands.

[15]In writing this case study, we recognize the need for more detailed research from gender perspectives on conflict and resistance to fully understand the impacts on indigenous women. The research questions, among others, include "How is the continuing land conflict changing the historically egalitarian Iban *Adat*?" "How do traditional societies disempower women—who could earlier be head of longhouses—when gender and other power relationships change with new land tenure systems?" and "Is the egalitarian system altered after lands are returned to communities by the judiciary?"

now at least, is an encouraging sign for indigenous communities to seek legal support against the might of the State and private companies. However, filing land dispute cases in courts has not always been easy for indigenous communities in Sarawak primarily because many do not have the resources needed. Realistically, trials can drag on for years; the Kpg. Lebor case was decided 14 years after the date of filing, and the process still continues with the Federal Court considering an appeal. The court process is costly even with *pro bono* lawyers, due to the high costs of travelling, food, and, at times, accommodation, when attending court trials (Wee 2012). Furthermore, attending courts means neglecting livelihood activities like farming in the village. While only certain people were expected to attend court hearings, other villagers had to assume their responsibilities. For example, women and older girls assumed the burden of farming and looking after the children and the elderly when the men went to court. Women also provided mental support to their men who were under constant stress or pressure from the government and companies, when they themselves and their children were also targeted. Money for household needs was markedly reduced as case-related expenses had to be met.

Native landowners once had gender-equal access to lands and resources. Today, land is increasingly controlled by the State and private companies that are male-dominated. As a consequence, native landowners are faced with growing landlessness and poverty, with poorer families and women more affected. The loss of NCR lands can result in the loss of indigenous resource management and local knowledge systems, by shifting access and control rights from community and family hands to political and economic elite. Through their everyday contact with and dependence on natural resources, women acquire knowledge about nature and their surroundings. The loss of land can cause women's roles and functions in traditional rituals and customs to disappear and transform the food self-sufficiency of communities to food dependency. Land privatization and land-use changes also affect women's rights to access the land and produce food for themselves and their families.

The practices of the people of Kpg. Lebor provided an insightful contrast to modern land-use patterns that seldom devote adequate attention to human and environmental impacts. Commercial logging and large-scale conversion of lands into commercial oil palm and other tree plantations development—in which greed and corruption of local political and economic elites, and the excessive consumption of the industrialized states are two of the underlying causes—are not only depleting natural forest resources but also significantly contributing to forest loss and environmental degradation. Consequences of environmental degradation and loss of forests and forests resources impact most on the rural poor and indigenous communities. Their vulnerability include diminishing lands for food production, scarcity of food from the forests, rivers and other natural resources, changing weather and climate patterns (severe drought, extreme rains, flooding, etc.), high rates of biodiversity loss (e.g. extinction or loss of rare/endemic flora and fauna species), degradation of ecosystems, and so on. Lack of access to NCR and the degradation of NCR lands has not only deprived women of access to forest land, but also access to clean water.

Furthermore, monoculture cash crops such as oil palm use significant amounts of chemical fertilizers and pesticides, contributing to pollution and other environmental hazards or rivers contaminated by pesticide run-off. Worse, the open burning of peatland forests for conversion to large oil palm plantations can lead to smouldering fires that linger for days which, in turn, can contribute to increased haze.[16] But in many cases, the Malaysian government officials and plantation companies often point fingers to swidden cultivators as the main culprits.

Environmental degradation increases vulnerability to disasters such as flash floods, mudslides, flooding, and extreme droughts. Poor people and especially women within groups that are already marginalized, such as indigenous peoples, often bear the brunt of disasters connected to increasingly degraded environments and resource scarcity. For the Kpg. Lebor people, therefore, winning back their lands goes beyond a legal victory. The challenge now is to strengthen customary rights to land and promote the interests of the community as a whole while ensuring gender-equal opportunities and rights in local resource management and in building resilience against man-made disasters.

2.11 Conclusion

In Kampong Lebor, people's relationship with land has changed dramatically over the past 51 years. Beginning with *usufruct* rights shared by an individual family within large community lands, we saw State takeover and privatization for large-scale oil palm planting in late 1990s. Then the legal struggle, ending with the return of a fraction of the disputed NCR lands to the Lebor people. The nature of land use has changed from family-worked subsistence agriculture to community-owned plantation bringing in a cash income in just two generations.

The implications for women are more far-reaching when the entire community is faced with insecurity over the tenure of their NCR lands. Women were more affected because of their higher dependence on forests for subsistence food, fuel and water. Though the courts have restored people's rights over land, the decision is being appealed. Further, the plantation is not being uprooted to restore the forest. Instead, the community is harvesting oil palms for income generation. This income is being handed to men, while women have lost their access to forest-based resources.

Women's unequal opportunities to access rights and use land from within their own society is exacerbated when the application and practice of *adat* law has been undermined by the State. Men could use this to deny women the rights to access communal lands. Women may also lose their access and control of specific forest resources that are sacred to spiritual activities, when such resources have

[16]The role of Malaysian and Singaporean companies in forest burning has been documented. For example, a police investigation into three Wilmar subsidiaries in Landak, West Kalimantan, found that they "were guilty of burning land intentionally and systematically with the intent to clear land for plantation development" (Milieudefensie, Lembaga Gemawan, & KONTAK Rakyat Borneo, 2007: 26 cited in Pye 2009).

degraded. For instance, Iban women's role as the guardians of the *padi pun*, or padi spirits, would be diminished or eliminated. With the new land use patterns, men have gained control over cash income, making them dominant within the family as well. This in turn will negatively impact on women's role and position in the family and community. The next generation of girls will likely have few resources and little control over their own lives. The degradation of forest and environment also affected people's access to clean water. Access to clean water is far more important for women than for men, since women are often in charge of washing and cooking.

There is a need to ensure that women's roles in the struggle for indigenous lands are openly acknowledged. As people adapt to live within this man-made disaster, it is critical that the new social structure be kinder to women and girls and ensure that they have access to and control over natural and financial resources to make their lives less difficult in the long run.

References

Bryan, Jane E.; Shearman, Philip L.; Asner, Gregory P.; Knapp, David E.; Aoro, Geraldine; Lokes, Barbara, 2013: "Extreme Differences in Forest Degradation in Borneo: Comparing Practices in Sarawak, Sabah, and Brunei", in: *PLoS ONE*, 8,7: e69679. doi:10.1371/journal.pone.0069679.

Census, 2010: *Population and Housing Census of Malaysia (Mukim Sarawak), 2010.* (Malaysia: Department of Statistics); accessed online (21 October 2013).

Colchester, Marcus; Wee, Aik Pang; Wong, Meng Chuo; Jalong, Thomas, 2007: *Land Is Life: Land Rights and Oil Palm Development in Sarawak* (Moreton-in-Marsh: Forest Peoples Programme—Bogor: Sawit Watch).

Hansen, Matt. C.; Potapov, P.V.; Moore, R.; Hancher, M.; Turubanova, S.A.; Tyukavina, A.; Thau, D.; Stehman, S.V.; Goetz, S.J.; Loveland, T.R.; Kommareddy, A.; Egorov, A.; Chini, L.; Justice, C.O.; Townshend, J.R.G., 2013: "High-Resolution Global Maps of 21st-Century Forest Cover Change", in: *Science*, 342,6160: 850–853 (15 November).

IDEAL, 1999: *Tanah Penghidup Kitai—Our Land is Our Livelihood* (Sibu: Ideal).

Matsubara, Tomomi, 2003: *Society and the Land: Contemporary Iban Society, Development Policy, and the Value of Native Customary Rights Land in Sarawak, Malaysia* (Unpublished thesis; Kota Samarahan: Institute of East Asian Studies, Universiti Malaysia Sarawak).

Pye, Oliver, 2009: "Palm Oil as a Transnational Crisis in South-East Asia", in: *Aktuelle Südostasienforschung/Current Research on South-East Asia, ASEAS*, 2,2: 81–101; at: http://www.seas.at/aseas/2_2/ASEAS_2_2_A5.pdf (2 August 2014).

Ross, Michael L., 2001: *Timber Booms and Institutional Breakdown in Southeast Asia* (Cambridge: Cambridge University Press).

SACCESS, 2008: *Adat and Human Rights in Sarawak*, Human Rights Report, SUARAM, launched on July 23, 2009.

Wee, Aik Pang, 2012: *Guidebook on Reclaiming Sarawak NCR Lands in Courts: Practical Information for Communities on Resorting to the Court Process (Civil Litigation) to Reclaim Native Customary Rights (NCR) Lands* (Kuching: SACCESS).

Yong, Carol, 2010: *Logging in Sarawak and the Rights of Sarawak's Indigenous Communities* (Brussels–Moreton-in-Marsh: Fern); at: http://www.fern.org; http://hornbillunleashed.wordpress.com/.

Chapter 3
Displacing Women, Resettling Families: Impact of Landslides on Women's Land Tenure Rights in Sri Lanka

Subhangi M.K. Herath

Abstract Landslides occur regularly in the hilly areas of Sri Lanka. This chapter focuses on the proposed voluntary displacement and resettlement process that took place in a landslide-affected hilly town in Sri Lanka. A gender impact assessment of the process reveals two major effects on the lives of women: it deprived them of economic opportunities since they would be resettled far away from the original area which offered them livelihood; and it affected land ownership rights accorded to women by traditional legal systems. Considering that the unit of analysis for resettlement was the household, patriarchal bias in official decision-making tended to confer ownership of the new property on the official head of the household, often a male, even if the de facto owner of the original property was a woman. The coexistence of tradition and modernity in a changing social environment created contradictions among women and men. Trapped between these paradoxes, women faced the threat of increased vulnerabilities and erosion of their traditional rights.

Keywords Land tenure · Women · Disaster · Social change · Landslides

Abbreviations

NBRO National Building Research organization

3.1 Introduction

The impact of disasters on human life and society has become a global concern with the increased intensity and frequency of natural and human-made disasters globally. Events such as the Indian Ocean tsunami of 2004, the Haiti earthquake of 2010, and the Fukushima disaster of 2011 have altered the global view on disasters, and disaster

---

Subhangi Herath, Department of Sociology, University of Colombo, Colombo 03, Sri Lanka. Email: subhangiherath40@yahoo.com.

K. Kusakabe et al. (eds.), *Gender and Land Tenure in the Context of Disaster in Asia*, SpringerBriefs in Environment, Security, Development and Peace, DOI 10.1007/978-3-319-16616-2_3

management has become a universal priority due to the huge impact on the lives of individuals, communities, and societies. However, these impacts differ across demographic and social groups: rich and poor, young and old, and women and men. The gendered impact of disasters has been a major concern in recent years not only due to the identification of the differential impact based on biological differences but also, and primarily, due to the socio-cultural positioning of women and men.

Social class and gender are the most crucial factors in any discussion on the problems faced by communities displaced due to natural disasters. Although development projects have also created large-scale internal displacements, unlike in the war and natural disasters, the planned development projects have included resettlement plans to minimizing the effect on the evacuees (Herath/Lešnik 2008). However, depending on their social position and capability, the affected communities or individuals wield varying power over displacement and resettlement decisions, perhaps diverting project locations towards vulnerable communities, increasing their vulnerabilities (Herath/Lešnik 2008). Disasters reduce the decision-making power of underprivileged communities and individuals due to the urgency of the situation.

From a gender perspective, it is obvious that women have considerably low decision-making power. The decision making process seem to have entirely neglected the idea that comprehensible disaster risk management during natural or socio-natural disasters have to be particularly concerned about differential and unique experiences of women and men during its different phases (Garcia 2005). More women than men lost their lives to the tsunami and women were discriminated against and harassed during the recovery process (Government of Sri Lanka 2005; Human Rights Commission 2005). This chapter adopts a gender perspective to examine issues concerning the resettlement of a community at risk, attempting to understand how gender differences make one group more vulnerable than the other.

3.2 Landslides in Sri Lanka

In recent times, landslides have become a grave national disaster in Sri Lanka due to their frequency. The country's central hills are particularly prone to landslides due to their geographical and geological composition. This has severely affected life, vegetation, and property in the area. According to information available at the *National Building Research Organization* (NBRO), the main government agency involved in landslide research in the country, the risk of losses due to landslides is much greater in urban and semi-urban locations due to population density, concentration of economic activities, and infrastructure facilities in such areas. NBRO records for the past 25 years show that not many disastrous landslides were reported during and before the first half of the twentieth century with three lives reportedly lost in two fatal landslides out of a total of nine incidents reported. However, a gradual increase in the occurrence of landslides was reported within the third quarter of the century. Of the 17 reported landslides during this period, two caused

44 fatalities. The last quarter of the 20th century saw a sharp escalation, making landslides an annual occurrence. Of the 75 recorded landslides during this period, 26 were fatal, causing 108 deaths, rendering thousands homeless and resulting in major economic losses. A sudden increase in the occurrence of landslides was observed from 2003 to 2007, resulting in severe loss to property and human life (Table 3.1).

One of the most obvious reasons for increasing landslides is the excessive human intervention in fragile natural hill slopes. Ten landslide prone districts have been identified in Sri Lanka, including Kandy where the study was conducted. The rapid increase in the occurrence of disastrous landslides compelled the Sri Lankan government to introduce a cabinet paper addressing the issue (Cabinet paper 116 of 16 June 1986). The cabinet then decided to initiate and launch "Landslide Hazards Zonation Mapping" (Ambalavanar/Lankanesan 1994). The NBRO was appointed the executing agency for this project (ibid.) which was extended to other landslide-prone areas in 1995 by a second cabinet paper (Cabinet paper 03/1372/111/061, dated July 17, 2003) (Arambepola/Weerasinghe 1998).

The study was conducted in two phases. An initial study on the socio-economic situation and the livelihood vulnerability in the landslide-affected township of Nawalapitiya in Kandy District of the Central Province of Sri Lanka was conducted in 2000 as part of a disaster risk-reduction project launched by NBRO, which interviewed 300 households. It was a broad socio-economic study conducted in two urban landslide-prone locations: one pre-disaster and one post-disaster (resettlement had already been proposed for these locations). The issue of land tenure was not a major concern, but livelihood issues and the potential for resettlement were directly addressed. The landslide-prone locations were revisited in 2010 to investigate the vulnerability of women facing disasters, as part of the author's ongoing academic research. In-depth interviews were conducted with a purposive sample of 15

Table 3.1 Occurrence of landslides in Sri Lanka 2000–2010

Year	No. of people affected	No. of lives lost	Houses damaged and destroyed	Damage to paddyand other crop land (Ha)	Relief distribution (Rs)
2000	23	0	4	2	60,000
2001	10	3	2	0	0
2002	2299	12	76	55	2,657,593
2003	22,338	218	3713	80	152,000,000
2004	3867	8	548	3	951,523
2005	1613	4	107	4	966,788
2006	26,889	38	2283	46	10,047,180
2007	27,497	34	2317	5713	3,167,719
2008	3180	19	283	11	3,283,260
2009	1376	11	117	4	712,430
2010	833	4	18	0	0

Note 1US$ = 130 Sri Lankan Rupees

Source Data were compiled from the database at Desinventar.lk on 06/11/2013 by the author

families who were considering moving out of the disaster prone locations. Case studies were conducted to examine various aspects of the vulnerability of women affected by landslide-related displacement: the position of women and their decision-making power within the family and the community; property ownership among women; their access to resources; their responsibilities, obligations and the impact of these on their lives; and the cumulative effect of all this on the capacities, aspirations, and lives of women.

When the area was revisited, many families were found still living there although many disaster-mitigation activities had taken place since 2001. This chapter is based mainly on three qualitative case studies in which women's land tenure rights had been discernibly affected also with reference to other case studies and interview narratives of other women, and it attempts to delineate diverse aspects of the land issue encountered by women facing disasters.

3.3 Study Area: Nawalapitiya

The displacement and resettlement process in a pre-landslide and a post-landslide site in Nawalapitiya are studied here. The process was similar to resettlement procedures adopted during development projects since this was a planned resettlement rather than a sudden evacuation. Once the area was identified as landslide-prone during NBRO's landslide-mapping process, awareness raising and mitigation activities began. People were requested to consider evacuation and resettlement as a prevention and mitigation measure.

The mitigation process in Nawalapitiya was considered successful due to several reasons—early identification, early planning of prevention and mitigation measures, increased public awareness and action taken towards mitigation, commitment of the authorities, and the personal interest and commitment of the Chairman and officials of the local government authority (Senanayake 2003). The Chairman was trained in disaster mitigation and addressed the issue on the basis of his knowledge and training. Despite this, certain actions of the local authority such as improper solid waste and sewage disposal and failure to curb encroachment and illegal construction potentially contributed to landslides. From personal experience, the people living in the study locations in Nawalapitiya knew the considerable damage landslides could cause. High annual rainfall and precipitous hills together with unplanned, environmentally destructive human activities including excessive encroachment had made the area landslide-prone.

The post-landslide site was a low-income community where women and men provided unskilled labour in the urban council or other urban activities. Women had fewer employment opportunities compared to men although both men and women faced a dearth of paid work in the area, nevertheless, proximity to the city offered women some alternatives in the informal sector. It was a traditional community in which people, especially women, had ownership of the lands they occupied. In contrast, most families in the pre-landslide community lived on encroached land;

therefore, few of them had legal ownership of land although they were economically better off than the post-landslide community. Since the area was located close to the central town, education and employment levels were better in the pre-landslide community. However, there was a greater threat of rock falls and landslides.

Nevertheless, both communities were reluctant to move to new locations during the voluntary resettlement that gave them the power to decide, along with an awareness of potential threats. Despite the differences in their socio-economic status, people in both locations lived in constant fear. Any sound had the potential to trigger a panic attack. One reason people refused to move was the failure of authorities to offer reasonable resettlement options. The only location suggested was approximately 15 kms away from the town. The likely negative effect that such resettlement could have on livelihoods was why people chose to live with the danger instead of moving.

Scholars on gender, disaster, and environment felt that owing to their close links with the environment women were knowledgeable about disaster risks (Mies/Shiva 1993; Kottegoda 2001; Ariyabandu/Wickramasinghe 2003). However, this view is overturned by the situation in Nawalapitiya. Women were aware that the area they lived in was prone to landslides but they did not take the situation seriously. Living with landslide risks for years without any life-threatening experiences to unsettle them had led them to underestimate the risks. A woman in the pre-disaster community stated:

> Those rocks were there during my childhood and my mother's and her mother's time just like they are now. This is the land I got from them. I am not ready to leave this for nothing. With God's will, many more generations will live here peacefully.

She was unconvinced by the scientific findings of the hazard-mapping process or the environmental changes visible in the area and simply did not want to worry about it. Women in these locations were at the greatest risk since most of them were housewives who stayed at home while men travelled out of the village for work. Damage and destruction of the property and having to separate from lands and houses to which their lives were intimately and sentimentally connected could exert incredible pressure on women. While these sentiments cannot be denied, this unflinching observance of tradition could also be a gendered outcome of women being constrained within a conservative perimeter that limits their mobility, access to education, and other social forms of knowledge, although the environment they live in is changing rapidly.

People in the post-disaster low-income community were more inclined to consider resettling, rather than live in constant fear and risk their lives. However, it was not considered an option by the pre-disaster community. That the new location would also have an impact on their class situation influenced their decision.[1] Land

---

[1]Studies have shown that low-income groups suffer most in disaster situations as they lose the little they have and are left with basically nothing to rebuild their lives and women of these underprivileged groups suffer the most (Lohani/Acharya 2003; Ariyabandu/Wickramasinghe 2003). Lack of decision-making power in the resettlement processes also exposes the lower classes and women to numerous negative ramifications.

ownership was given less importance in the pre-disaster location due to the high prevalence of encroached land in the area. Men, specifically, laid great stress on livelihood which was directly connected to their proximity to the urban area.

The study revealed that the risks faced by women cannot be understood merely by acknowledging the gravity of the disaster. Social class, gender, traditions, customs, and culture further complicate the issue. The gendered nature of land tenure in the area was noticeably shaped by prevalent marriage customs and land-tenure patterns. The area belonged to the former Kandyan Kingdom, and land tenure and marriage norms came under the specific legal system called Kandyan Law. After the subjugation of the country by the Portuguese, the Dutch, and subsequently the British, colonial legal systems[2]—the Roman-Dutch law on land tenure and the British Common Law on registration of marriage—came into practice in the area. Two other ethnic legal systems governed land tenure simultaneously: the Thesawalamai Law of Tamil Hindu population and the Shariah Law of the Muslims. The local people followed these different legal systems according to their ethnic, religious, and cultural preferences. Although the commonly accepted legal systems among the majority Sinhalese community[3] are the Roman-Dutch Law and the British Common Law, in Kandyan areas, people still have the option of marrying according to Kandyan Law which can also be applied to their property rights.

Most people participating in this study followed either the Common Law or the Kandyan law. The latter accorded women some exceptional rights with respect to marriage, divorce, and property ownership.[4] According to the Kandyan marriage and divorce law, property belonging to women at the time of marriage (movable or immovable) remained their personal property even after marriage. Their spouses did not receive the legal right to utilize their property for personal needs or control their property after marriage.[5] The case studies revealed that people's decisions on how to handle disasters were clearly influenced by family customs and the legal systems they followed. Women's experiences demonstrated that land tenure was part and parcel of this highly complex situation.

---

[2]Then known as Ceylon, the country was under European rule for almost five decades. First, the coastal areas fell under Portuguese rule, then under Dutch rule, and finally under British rule. The Kandyan Kingdom was subjugated by the British in 1815.

[3]Sinhalese form the majority (74 %) in Sri Lanka followed by Tamils (16 %) and Muslims (8 %) (Sri Lanka Census of population and Housing 2011).

[4]The Government of Sri Lanka, Marriage and Divorce Act (Kandyan) Act Nos. 44 of 1952; 34 of 1954; 22 of 1955, Chapter 132, August 1, 1954 (Act 41 of 1975); Par 4 Section 32—Divorces.

[5]The Government of Sri Lanka, Marriage and Divorce Act (Kandyan) Act Nos. 44 of 1952; 34 of 1954; 22 of 1955, Chapter 132, August 1, 1954 (Act 41 of 1975); Par 4 Sections 8 and 19.

3.4 Feminizing Risks

Srimathie (40) has studied up to Grade 8. Her marriage to Sunil was arranged by the family. The couple has three school-going children, aged 10, 12, and 16. She is a housewife who grew vegetables in her garden for her family's consumption. Her husband is a labourer in the permanent cadre of the local government authority and earned an income adequate for their survival. She was given a block of land which her mother had inherited. Following are extracts from discussions with her on the land issue.

> When the officials came to ask us to move I was adamant that I would not leave. The whole village was together in the decision. Some people said that the disaster would not happen. I was a bit sceptical too, but after attending the meetings we were very scared. None of us liked the land suggested by the authorities. We were also not sure if we would get it or not. After a couple of years, landslides became an annual event. We were so scared. I was worried about my children. I never had a happy life. My husband had the habit of putting me down whenever he got a chance. He used the official claim that this was a high-risk area to condemn the land I had received as dowry. My mother had got it from her mother so I was very proud of it. We did not marry according to the Kandyan law as my mother did, but my mother insisted that the property should go to all her children. We were five in the family. My sister and I got blocks from this land and our three brothers got the *mahagedara watta* (main land).
>
> My husband and I had constant discussions and arguments about moving. I was very sure that he would talk to the authorities and use his connections to get the new land under his name. I did not dare to say anything against that. I did not want to lose my family because of land. If he leaves me and my children, who will support us? They should have given the land to the person who owned it instead of asking 'The husband is the head of the household, isn't he?' I said, 'Yes, but the land is under my name'. My husband was furious that I had claimed the right to the land. The next morning he asked me if I still wanted to stay and risk the lives of our children. So I said I would like to move. I was also in constant fear for the children's safety. After about a year, we moved to the new location. Travelling to his job is now very difficult for him but I think he enjoys living in his own place.

This case illustrates that vulnerability cannot be examined merely in the context of natural hazards; it must be linked to social, cultural, economic, political, institutional, physical, and environmental factors (Herath 2003). In the Mahaweli River Diversion Project, one of the largest development projects in the country, where new land was given to the legal owner of the evacuated land, almost half the resettled land was owned by women.[6] However, there is no evidence, quantitative or qualitative, regarding gendered land ownership in case of disaster-related resettlement. Case studies show that disasters are unique situations where normal rules and social processes do not apply. Among the many 'unusual' events taking place during and after disasters is women's loss of traditional land rights.

Social changes include loss of privileges enjoyed by women in traditional societies, including land tenure and inheritance (Tripp 2004). The traditional inheritance and marital law of Sri Lanka or the 'Kandyan law' accorded significant

[6]Field data from the Mahaweli System C resettlement area.

property rights to women. Although fewer marriages now take place under the Kandyan law, people customarily follow some of these laws in their practices. The case of Srimathie delineates how 'modernity' or the "post-traditional society" and the risks and insecurities created by it could lead to the collapse of the traditional social system and the privileges it accorded to women in contrast with how modernity is perceived as influencing societies in the West (Giddens 1990; Beck 1992). While the environmental risks 'manufactured' (Giddens 1999) by the urbanization process and the form of governmentality (Foucault 1980) produced by scientific expert systems of knowledge, state policies, bureaucracy and patriarchy upsurge the male dominance in the public sphere, women become pushed back to their own powerlessness within the private sphere, depriving them of their traditional land tenure rights which was women's main source of power in traditional societies (Shiva 1988; Agarwal 1998; Chowdhry, undated). In order to preserve the institution of family in confrontation with modernity, women forego their traditional autonomy and bargaining power along with their land rights making them vulnerable to increasing domestic risks that are intertwined with the influence of global markets and technologies (Denney 2005).

3.5 Normalizing Patriarchy

Dayani (28) was married recently and has one child. Her husband is employed as a casual worker in a shop in the town. She has a brother who works abroad as a driver. She lives in her parental house which was legally given to her and which now falls within the demarcated disaster-prone area. She has no source of income other than land ownership. The home garden is maintained by the entire family and supplements the income provided by her husband. Her father is a retired government employee who worked as a labourer. He spends his entire pension on the extended family which is a major support for them. They were still considering relocating at the time of the study.

> For over 5 years now, we have been debating on whether to leave this place or not. This is a sin. We can never have a proper life. During the rainy season, I hardly sleep. Even before marriage, I remember my parents and I would be awake at night during heavy rains. Awareness programs have made us more scared about the situation. We live right under the rock. My mother says that when she came here after marriage, she heard stories about landslides from her mother-in-law. She says that we don't have to worry so much because there are signs that become visible long before a landslide occurs, but the officials say that landslides could occur without adequate warnings. There are cracks in some houses but not in ours.
>
> How can we move without a proper location? They (officials) don't give a concrete answer when we ask about this. This is the house we have lived in for three generations. My parents gave the house to me because I am the only daughter and they expected that I would take care of them in their old age. So, if I move, I have to take them with me.
>
> I am really worried because of something my father did. When the officials came to survey the houses, my father gave my husband's name as the head of the household without consulting any of us. Both my mother and I were very upset about it. He asked me if I could

> run around like a man for official matters when it comes to resettlement. But this decision of my father worries me a lot. People say that the new land will be given to the head of the household and not to the real owner. My husband is a good man, but people change when it comes to money and property matters. I am not worried about myself. I am worried about my parents and brother.
> We may have to take a decision at any point. If it goes bad, I think it is my destiny to lose everything I have. I think it would be as if a curse fell on us if we had to leave the house in which three generations of the family have lived. The whole thing is a curse. How many families are living in fear?

People, except a few, did not view the landslide disaster as a 'normal' situation. The more they saw it as 'unacceptable' the greater was the difficulty for them to cope with the situation. Nevertheless, that tension and uncertainty marked their handling of the situation with caution. The woman saw it as a curse on her to deprive her of everything she owned and therefore, she was still reluctant to move. This is a unique situation where the father took a drastic property-based decision which was not in his daughter's favour, but was based on his patriarchal ideologies and his hopes of protecting the daughter. She was well aware of the dangers posed by that decision, which made it difficult for her to move forward. She was vacillating between relocating and not relocating, trying to conform to the gendered expectations of the society and continuing to live with the risk. Men, on the other hand, did not problematize the issue of women's land rights or their subordinate position in society. They believed that it is best for the family to have land titles in the name of the man, who is seen as the natural head of the household.

Dayani's account shows that there is considerable danger in seeing disasters as "part and parcel of the 'normal' development of societies—as unresolved problems of development" (Ariyabandu/Wickramasinghe 2003: 31). Such a view puts all other related aspects within the realm of 'normality'. Neither disasters nor other socially constructed situations could be viewed as 'normal'.[7] Both disasters and gender inequalities are intertwined 'abnormal' situations that have been normalized in society. As a result of this, the gendered impact of natural hazards is also normalized, preventing its identification and rectification.

The losses suffered by women due to resettlement varied across the communities studied. Women in the post-disaster community were more vulnerable because: (1) more women owned and inherited land; and (2) women had a much lower social status with no power to negotiate. Working women in the pre-disaster community were economically better off, yet many of them did not own land so they were less likely to lose personally in case of resettlement. This is an obvious dilemma faced by women in vulnerable situations. Although economic power bestowed by property ownership and livelihood activities gives women greater bargaining

[7]Most natural disasters have a human/social stake and, therefore, can often be prevented or mitigated if proper actions are taken. According to Strydom (2002), contemporary risks are not simply physical or objective properties of a real phenomenon but are social or collective constructs.

power, the socio-cultural construct within which they live could make such women more vulnerable to losses during disasters.

Resettlement usually takes a top-down approach. The authorities are chiefly concerned with identifying potential resettlement locations and settling affected people. Typically, the theme underlying resettlement is "land for land" and "house for a house". Often, both women and men in affected communities have no power over the decisions taken by the authorities; however, due to existing gender inequalities, women become almost invisible in the decision-making process. Women are often not seen as important stakeholders during community-level decision-making (Enarson et al. 2007). During the author's study in tsunami-affected communities on the southern coast of Sri Lanka, a woman stated:

> When they [authorities] come for discussions, they do not pay attention to us. We women know what we want for the family, for our children. When they were ignoring us, I raised my voice. I did it for all of us because otherwise the families without men would have to face all the disadvantages since they [men] have all the connections, not us. But later they told people that nothing can be done with that crazy woman [that is me].

Women heads of households often did not have any say in resettlement. They had to accept what they were offered for two reasons: (1) the authorities and the community considered them the most vulnerable, so attempting to make choices was seen as inviting trouble and ending up having nothing and (2) women perceived themselves as weak and helpless in the absence of a man in the family, so they felt compelled to accept whatever was offered to them. The following narrative by a woman from a tsunami-affected village clearly depicts the level of vulnerability some women faced:

> People went to Kachchery [divisional secretariat] almost every day. I could not do that due to my busy schedule. On the day of the meeting, I did not accept the house that was given, because I did not want to raise my children in that surrounding. People asked me not to turn down the offer, but I did. During the second meeting, I was given a place which was a little better but very far from the rest of my family. I do not know why only women with no husbands were treated like that. When I told them about my grievance, they scolded me. They said that women in my situation should take whatever is offered. Finally, I accepted the land. There were numerous rumours about the unfair practices taking place during the land-allocation process but I do not know if they are true.

Livelihood vulnerability has been discussed at greater length in Sri Lankan literature that deals with the gender aspects of disaster management (Fernando/Fernando 1997; Gomez 2006; Kottegoda 2001; Krishnaswami/Dave 2001; Ariyabandu/Wickramasinghe 2003; Ariyabandu 2003). However, the Nawalapitiya study indicates the need to view gender-based vulnerability as a 'complex totality' that encompasses all aspects of social life rather than a phenomenon produced by a single cause. Interpreting gendered vulnerability as an outcome of disaster alone creates the possibility of debilitating the disaster-management process, especially rehabilitation and resettlement (Herath 2003, 2006). As "Guidelines for Gender Sensitive Disaster Management" (APWLD 2006a) points out, social analysis of disasters must ensure that gender discrimination encountered by diverse groups of women in different situations is addressed.

3.6 Neglecting Women's Knowledge

After her marriage, Samanmalee (35) inherited a small block of land from the family compound. Married at 20, she is now the mother of two children. Her husband works as a casual labourer. Since her husband's income is insufficient to meet the family's needs, she joined a tailoring shop as a seamstress.

> I could go to work because we lived in the same compound as my parents. Even after the area was identified as disaster-prone, we did not consider moving until we noticed a crack on the wall of the house. Then we decided that it was not a good idea to live there anymore. My parents still live there, but they often stay with my siblings, so nobody really lives there permanently and I am glad about it. I feel sorry about the land. We cannot sell it either because of the disaster situation.
>
> We had many discussions with officials but they did not come up with a fair solution. Like us, many people work in the town, so how could we move to a faraway place? At the meetings, I asked many questions which they did not have answers for. They were good at explaining the situation and making us feel that we should leave the place immediately, but they did not give us a reasonable place to settle down. I know the area well and I suggested possible places for resettlement, but they found some problem with all of them. Only a few women spoke. When I spoke about garbage dumping and other problems we faced, they did not pay much attention. The chairman is very good but there were others who did not like women speaking up. They directed their questions to the men. When they came to my house, they did not ask anything but included my husband's name as the owner of the house. I did not have a problem with that, but if my father had known he would have been very unhappy. There are many 'binna'[8] marriages in this area and they should have known that at least.
>
> We were fortunate that we got a small block of land from a gentleman under whom my husband worked earlier. We first built a small hut there and then made this room with cement blocks. When we decided to move, the gentleman asked whom he should transfer the land to. He [the gentleman] knew that I owned the previous land. I suggested his [husband's] name. I had already included my husband's name as the head of the household; otherwise, it would not have been good for family harmony. I know he feels good about it. He is the man and he should own the land. On the other hand, how could I say that I need the land? It is because of my husband that the gentleman gave us the land.

The much-discussed issue of 'familism'[9] as a sexist bias in social research (Eichler 1991) seemed highly applicable in the official decision-making process in this case study. Women's concerns about disasters and their traditional knowledge could be considered a capacity in disaster preparedness (Ariyabandu et al. 2003), yet Samanmalee's case demonstrated that top-down approaches adopted in prevention and mitigation programs seriously neglected women's capacities. Such negligence of women in decision-making was mainly based on derogatory societal

---

[8]'Binna' is a traditional system of marriage that came within the purview of the Kandyan law where a man moves into a woman's property after marriage and the woman is accorded considerable rights.

[9]Eichler (1991) points out that considering family/household as the unit of analysis could overlook the fact that women and men living in these units could enjoy vastly different opportunities and treatments.

attitudes towards women and their knowledge, thus making women an underutilized or an unutilized resource (Masood/Wahra 1997; Ariyabandu 1999).

The case also demonstrated the contradictions in women's social existence. Despite her confidence and knowledge, Samanmalee was ready to sacrifice her land entitlement to her husband due to her perceptions of harmony in a marital relationship, which were, in turn, shaped by her conservative upbringing.

3.7 Conclusion

In Nawalapitiya, all three women were ready to forego their land rights: (1) they had to submit to patriarchal domination, however discriminatory, to ensure the economic and social stability of their families[10] and (2) the gendered cultural ethos persuaded women to unquestioningly accept male authority (even unwillingly), even if it meant relinquishing rights and privileges they valued and felt intimately connected to in their lives. 'Male supremacy' was naturalized by women to justify their unwilling submission to what they called 'destiny'. Their entire rational social existence seemed to be ambiguous in a context where socialization and patriarchal values overruled women's capacity for reasoning, especially with respect to decision-making in the domestic sphere. Women's low gender sensitivity played a major role in paving the way for the loss of their land entitlements. The study demonstrated that the strategies adopted in disaster management were neither gender-specific nor receptive to women's knowledge, experiences and preferences. One of the reasons cited for this was men's domination of decision-making bodies, including local government authorities, and national politics.

Field data revealed that disasters cannot be considered 'normal' situations. Disasters could often be considered products of human activity and many 'unusual' events during and after disasters would change the 'normal' life of those affected. In the case of Nawalapitiya, a situation where women could lose their traditional land rights could be considered one such crucial event that could seriously disrupt the lives of women in the community. Although the economic power bestowed upon women by their property ownership and livelihood activities could be viewed as a strength that allows them greater bargaining power in a normal circumstance, the socio-cultural construct within which they live could make such women more vulnerable to losses during disasters. In highly stratified social contexts, the study questions the prevailing trend in relevant literature to 'normalize' many social or natural processes that lead to social change, including development, modernity, disaster, and gender. The study concludes that the risks produced by such phenomena could push traditional social systems to collapse, and with them, also the privileges they accord to women and other underprivileged groups.

[10]Compromises women make in the face of patriarchal power in order to safeguard the family unit have been well illustrated by several authors in different contexts (Ariyabandu/Wickramasinghe 2003; APWLD 2006b; Abeysekara 2005; Inform 2006; Rees/Bertolemei 2005; Action Aid 2007).

Faced with such precarious land ownership in the resettlement area, women prefer to stay in the original place, where they have more secure land ownership and livelihood. Their 'traditional' knowledge—based on the experiences of her grandmother and mother—says that the place is safe. However, women also understand that based on 'scientific' knowledge—what the government officers tell them—the place is not safe. So, they live in fear. Yet, in order to defend their entitlements, and with no alternative that would ensure their current livelihood, they use their 'traditional' knowledge to convince themselves. These cases show that understanding women's vulnerability to disasters solely from the perspective of natural disasters or solely from gender perspective could obscure the reality of this complex social phenomenon. Since gender relations vary within socio-historical contexts, social structures with their relevant social institutions, organizations, groups and relations must be examined to understand the gendered impact of disasters. Disaster literature repeatedly confirms that women are highly knowledgeable about disasters because of their close affinity with the environment, yet their social position makes them highly vulnerable to disasters (Jiggins 1994; Merchant 1996; Shiva 1998; Agarwal 1998; Nightingale 2006; World Bank undated). Drawing from both traditional and scientific knowledge, women understood the risks of living in a landslide-prone area. However, urgent concerns regarding their livelihood and family ties dictate their decision-making. In order to justify their decision to stay, they claim that nothing happened during their mothers' and grandmothers' time, ignoring the fact that the situation has changed since then.

The losses women faced due to resettlement were not equal across communities. The diversity of losses depended on factors such as male authority; level of female subordination; cultural ethos and values that governed attitudes and behaviour; socialization of women; and attitudes about women among women, their family members, and the society; and the official stance on the whole issue. Prioritizing women's needs in relief and rehabilitation work is crucial in reducing the vulnerability of women in disaster situations. The Nawalapitiya case demonstrated that this has to be done with a high level of gender sensitivity by incorporating women at all levels of disaster management, including preparedness, planning, prevention, and mitigation strategies. In addition, clear policies in support of women's land and property rights are required. This will help draft pro-women guidelines that specifically look at the socio-cultural positioning of women in each community and their corresponding gender responsibilities that extend beyond their own survival.

References

Abeysekera, Sunila, 2005: "The Gendered Nature of Natural Disaster", in: *Options*, 36,1.

Agarwal, Bina, 1998: "Disinherited Peasants, Disadvantaged Workers: a Gender Perspective on Land and Livelihood", in: *Economic and Political Weekly*, 33,13: A2–A14.

Action Aid, 2007: "Violence Against Women in the Post Tsunami Context—People's Report, Sri Lanka"—A report based on intensive discussions with 3,833 tsunami affected women in six districts in Sri Lanka (Colombo: Action Aid).

Ambalavanar, V.; Lankanesan, T., 1994: *National Disaster Management Plan—Current Status and Trends with Particular Reference to Landslide Management* (Colombo: Proceedings of the National Symposium on Landslides) 343–348.

APWLD, 2006a: "Asian Forum on Women in Disasters: Two Years after Tsunami Women Still Face Violations of Their Rights", in: Asia Pacific Forum on Women, Law and Development: *Forum News*, 19,3 (December 2006); www.apwld.org.

APWLD, 2006b: "Guidelines for Gender Sensitive Disaster Management", Chiang Mai: Asia Pacific Forum on Women, Law and Development.

Arambepola, N.M.S.I.; Weerasinghe, K.M., 1998: "Towards achieving the long term objectives of landslide hazard mapping programme", in: *Proceedings of the Workshop on the Role of R & D Institutions in Natural Disaster Management* (Colombo): 8-1–8-8.

Ariyabandu, Madhavi, 1999: *Defeating Disasters: Ideas for Action* (Colombo: A Duryog Nivaran Publication: Intermediate Technology Development Group).

Ariyabandu, Madhavi, 2003: "Women: The Risk Managers", in: *Voice of Women—A Sri Lankan Journal for Women's Liberation*, 6,1 (August).

Ariyabandu, Madhavi; Wickramasinghe, Maithree, 2003: *Gender Dimensions of Disaster Management* (Colombo: ITDG South Asia Publication).

Beck, Ulrich, 1992: *Risk Society: Towards a New Modernity* (London: Polity Press).

Chowdhry, Prem, undated: "Reduction of Violence Against Women: Property Ownership & Economic Independence in Rural Haryana", UN Women; at: http://www.unwomensouthasia.org/assets/Violence-Property-Rights2.pdf.

Denney, David, 2005: *Risk and Society* (London: Sage).

Eichler, Margrit, 1991: *Non-Sexist Research Methods: A Practical Guide* (New York: Routledge).

Enarson, Elaine; Fothergill, Alice; Peek, Lori, 2007: "Gender and Disaster: Foundations and Directions", in: Havidán, Rodríguez; Quarantelli, Enrico L.; Dynes, Russell R. (Eds.): *Handbook of Disaster Research* (New York: Springer).

Fernando, Priyanthi; Fernando, Vijitha, 1997: *South Asian Women: Facing Disasters, Securing Life* (Colombo: Duryog Nivaran Secretariat).

Foucault, Michel (ed. by C. Gordon), 1980: *Power/Knowledge: Selected Interviews and Other Writings 1972–1977* (Brighton: Harvester).

Garcia, Cecilia Castro, 2005: "Gender Inequality in the Comprehensible Disaster Risk Management—An Introduction", (Published in Spanish in): *Revista de la Cristobal Colon* (Veracruz), No 20, Tercera epoca, Ano iii (Translation by Emilia Reyes).

Giddens, Anthony, 1990: *The Consequences of Modernity* (Cambridge: Polity Press).

Giddens, Anthony, 1999: *Runaway World* (London: Profile Books).

Gomez, Shyamala, 2006: "Guidelines for Gender Sensitive Disaster Management"; at: www.apwld.org.

Government of Sri Lanka, 2005: *Sri Lanka: Post Tsunami Recovery and Reconstruction—Joint Report of the Government of Sri Lanka and Development Partners* (Colombo: Government of Sri Lanka, December).

Herath, Subhangi M.K., 2003: "Vulnerability in Trincomalee District", Unpublished Study Report submitted to Care International, Colombo.

Herath, Subhangi M.K., 2006: "Mobilising Women for Reconstruction in Sri Lanka (post tsunami)". Report of the evaluation of Kantha Shakthi Projects in Matara and Tangalle.

Herath, S.; Lešnik, B., 2008: "Internally displaced persons: Implications for the development of social work in Sri Lanka", in: Shula Ramon (Ed.): *Social work in the context of political conflict* (Birmingham: Venture Press) 77–97.

Human Rights Commission, 2005: Report of the Women's Division, Disaster Relief Monitoring Unit. Human Rights Commission in Sri Lanka (Colombo: Human Rights Commission, August).

INFORM, 2006: *Tsunami Aftermath: Violations of Women's Human Rights in Sri Lanka* (Colombo: INFORM Human Rights Documentation Centre).

Jiggins, Janice, 1994: *Changing the Boundaries: Women Centred Perspectives on Population and the Environment, Population and the Environment* (New York: Island Press).

Kottegoda, Sepali, 2001: *Livelihood Options for Disaster Risk Reduction in South Asia: A Study of Gender Aspects of Communities Living With Drought and Land Slides in Sri Lanka* (Colombo: ITDG).

Krishnaswami, P.B; Kumar, S.; Dave, M., 2001: *Gender Issues for Livelihood Options for Disaster Risk Reduction* (Ahmedabad: Disaster Mitigation Institute).

Lohani, Balkrishna; Acharya, Harish Chandra, 2003: "Impact on flood on gender: A case study", in: Sahni, Pradeep; Ariyabandu, Madhavi Malalgoda (Eds.): *Disaster Risk Reduction in South Asia* (New Delhi: Prentice-Hall of India Private Limited) 281–288.

Masood, Mahjabeen; Wahra, Gawher Nayeem, 1997: "Bangladesh: Learning to deal with disasters", in: Fernando, Priyanthi; Fernando, Vijitha (Eds.): *South Asian Women: Facing Disasters, Securing Life* (Colombo: A Duryog Nivaran Publication: Intermediate Technology Development group).

Merchant, Carolyn, 1996: *Feminism & Ecology* (New York: New York University Press).

Mies, Maria; Shiva, Vandana, 1993: *Ecofeminism* (Halifax: Fernwood Publications).

Nightingale, Andrea, 2006: "The nature of gender: work, gender and environment", in: *Environment and Planning D: Society and Space*, 24,2: 165–185.

Rees, Susan; Pittaway, Eileen; Bartolomei, Linda, 2005: "Waves of violence—Women in post-tsunami Sri Lanka", in: *The Australian Journal of Disaster and Trauma Studies*, 2005-2; at: http://www.massey.ac.nz/~trauma/issues/2005-2/rees.htm.

Senanayake, Dhanushi, 2003: "Timely Intervention for Disaster Risk Reduction". in: Sahni, Pradeep; Ariyabandu, Madhavi Malalgoda (Eds.): *Disaster Risk Reduction in South Asia* (New Delhi: Prentice-Hall India Private Limited): 250–259.

Shiva, Vandana, 1988: *Staying Alive* (London: Zed).

Sri Lanka Census of Population and Housing, 2011: Department of Census and Statistics Sri Lanka, at: http://www.statistics.gov.lk/PopHouSat/CPH2011/index.php?fileName=pop42&gp=Activities&tpl=3 (5 June 2014).

Strydom, Piet, 2002: *Risk, Environment and Society* (Philadelphia: Open University Press).

Tripp, Aili Mari, 2004: "Women's Movements, Customary Law, and Land Rights in Africa: The Case of Uganda", in: *African Studies Quarterly*, 7, 4: 1–19.

World Bank, undated: "GDRC Special Focus on Gender and the Environment", at: http://www.gdrc.org/gender/gender-envi.html.

Other Literature

Cabinet Paper 116, Government of Sri Lanka, 16 June 1986.

Cabinet Paper 03/1372/111/06, Government of Sri Lanka, 17 July 2003.

Government of Sri Lanka, 1952, 1954 & 1955: Marriage and Divorce Act (Kandyan).

Chapter 4
Impact of Flash Floods on a Matrilineal Society in West Sumatra, Indonesia

Yonariza and Mahdi

Abstract Following a flash flood in 2012 in a matrilineal community in Padang, West Sumatra, Indonesia, we examine the impact of the flood on women and men in their position as land owners and land users. Every disaster has a social impact that is larger than its physical impact. Land is a key resource for the livelihood in this community, and hence changes in land ownership have a long-term and deep-rooted impact on specific people who have lost or gained land rights. In matrilineal societies, women inherit land but their brothers are its guardians. Considering the matrilineal culture of the Minangkabau people, we also studied the gender and cultural sensitivity of the government and NGO responses to the disaster. The study used observation, key informant interviews, and secondary data. It showed links between gender, land tenure, and disaster in some cycles of disaster management, especially when damage was assessed and landowners identified. The study also reveals that women's inclusion in disaster preparedness is critical to make disaster management more gender equitable.

Keywords Flash flood · Padang · Gender · Land tenure · Matrilineal society

Abbreviations

BNPB	*Badan Nasional Penanggulangan Bencana* (National Disaster Management Agency)
GoI	Government of Indonesia
IDR	Indonesian Rupiah
NGO	Non-governmental organization
UN	United Nations

Yonariza, Professor, Forest Resources Management, Andalas University, Padang, Indonesia; Email: yonariza@gmail.com. Mahdi, Chair, Integrated Natural Resources Management field of Study at Graduate Program, Andalas University in Padang, Indonesia; Email: kuteihmahdi@gmail.com.

K. Kusakabe et al. (eds.), *Gender and Land Tenure in the Context of Disaster in Asia*, SpringerBriefs in Environment, Security, Development and Peace, DOI 10.1007/978-3-319-16616-2_4

4.1 Introduction

Located along the Ring of Fire, Indonesia is disaster-prone. Around 80 % of Indonesia's disasters are hydro-meteorological, i.e., caused by aspects of weather, such as floods, landslides, cyclones, and drought. In 2012, 50 natural disasters struck Indonesia (Mongabay Indonesia 2012) inflicting great losses of property and life.

The government of Indonesia (GoI) strengthened the National Disaster Management Agency or *Badan Nasional Penanggulangan Bencana* (BNPB) in 2008 to improve disaster management. As a national body, BNPB is a significant improvement over its predecessor. BNPB was created by a State law, Law No. 8/2007, that invests the agency with greater authority in disaster management and secures an operational budget for more responsive disaster management. Similar agencies were set up at provincial and district levels as well. This demonstrates a serious attempt in dealing with disasters.

According to BNPB, globally, natural disasters have increased 350 % over the past three decades. BNPB and its provincial and district equivalents need greater inputs for dealing with disasters, including awareness of land tenure and gender sensitivity.

4.2 Disasters in West Sumatra

The western part of Sumatra Island—along the Barisan Mountain Range—is a volcanic area with seven volcanoes, some of which are active. It is a Degree VIII earthquake-intensity area according to the UN Office for the Coordination of Humanitarian Affairs Regional Office for Asia Pacific (2011). In addition, the province faces the Indian Ocean and has the highest rainfall in Indonesia. With these landscape characteristics the province is prone to earthquakes, landslides, floods, and flash floods. In 2012 alone, 14 flash floods occurred in the province. Some of these disasters were unique in terms of magnitude and scale. The area also witnessed 25 forest fires, 13 landslides, six tropical cyclones, and two shore abrasions (West Sumatra Province Disaster Management Agency 2012).

The study was conducted in Padang City, Pauh sub-district, West Sumatra. The site is located in a valley near a river, where fast and dangerous flash floods can ravage farmland and settlements. This river, which was a source of water for irrigation, destroyed irrigation infrastructure with the flash flood and threatened local food availability. Paddy fields along the river were not protected and were buried or washed out when the river breached its banks. This unfortunate confluence of circumstances allowed for a study of the links between disasters, gender, and land tenure.

This chapter focuses on a flash flood although the area also witnessed many other disasters, including an earthquake in 2009 that recorded 7.2 on the Richter

Scale, and numerous landslides. Flash floods, a hydrometeorological disaster that affects land tenure in specific ways, are on the rise. Mongabay Indonesia (2012) stated that most floods were due to improper land management.

The chapter is based on a study conducted more than a year after the disaster although evidence of the disaster was still obvious. It uses the following data-gathering techniques: revisiting disaster sites for observation; interviewing key informants; interviewing women disaster victims; and analysing secondary data, news, and amateur videos and photos taken during and after the disaster. The study site is just below Andalas University campus where authors are affiliated, so they closely observed the calamity and its aftermath, supported with photos documentation. The study was intensified a year later with a brief survey, field observations, interview with important stakeholders including two leaders of farmers' groups, one head of settlement, an matrilineal guardian (*mamak*), six women farmers, a share cropper, a government official, and an agricultural extension worker. The authors also used triangulation technique to verify information collected from various sources.

4.3 Land Tenure in the Minangkabau Matrilineal Society

West Sumatra and some parts of Riau and Jambi provinces in Sumatra, Indonesia, are inhabited by the Minangkabau people, an ethnic group following the matrilineal system of inheritance. They are considered the largest matrilineal society in the world today (Kato 1978). Matrilineal tradition among Minangkabau has the following characteristics: descendants are organized according to the female line, thus, all children belong to the mother's clan or lineage; agricultural land, ancestral treasures, and miscellaneous *adat* titles communal properties; residential pattern is uxorilocal; and authority within a lineage is in the hands of the *mamak rumah* (mother's eldest brother) and not the father (Kato 1982: 51). Lineage systems dictate that ancestral property is only inherited by women. However, men supervise the utilization of this property. Minangkabau scholars believe that one can understand Minangkabau land tenure only by understanding their social organization (see Kato 1978; Naim 1984; von Benda-Beckmann 2004; Stivens 1996).

The Minangkabau matrilineal society is grouped into *samande, paruik, suku*, and *nagari*. *Samande* is the smallest unit and consists of a family with a great grandmother. *Paruik* is made up of several *mande*, while *suku* is a matriclan. *Nagari* is a territorial unit, a federation of *suku* or clan. This organization also reflects the land tenure system of the Minangkabau. Benda-Beckmann (2004) describes Minangkabau social organization as *pusako complex*, where they live in relatively autonomous villages (*nagari*) with common village land (*ulayat*). They are organized in localized matriclans, consisting of matrilineages headed by a *panghulu*. The council of the lineage heads of the founding matrilineages forms the village government. Villages are endogamous, post-marital residences being uxorilocal. In *adat* law and philosophy, the lineage members form a social, political, and economic unit. They are "one" in many respects. They are descended from "one womb." They share the leadership of one lineage head, the hereditary title of their lineage and their common inherited property, *harato pusako.* This lineage property-people complex is a unit in the economic and political constitution of Minangkabau villages.

Land tenure then is simply classified as *ulayat nagari* (which literally means common to a *nagari,* where all *nagari* members can access *nagari's* land) or *ulayat suku* (where members of a particular clan can access land), *ulayat kaum* (or common land of extended family), and *ganggam bauntuak* (land that has been granted to *samande*). *Sawahs* (irrigated land) are usually divided among *samande* under the supervision of the mother's eldest brother or *mamak*.

Most of agricultural land such as rice fields have been divided among *samande*, the smallest unit of a family with a common great grandmother, and the land is accessed by daughters under supervision of brother of grand mother (*mamak*). But the land is considered as belonging to the grandmother and is accessed by her daughters; the *mamak* is a figurehead.

Lineage land is shared among the *samande*, but they have only usufruct right over the land. The lineage land is under supervision of *mamak kaum* (head of lineage), but more and more lineage land has been divided among *samande* when they engage in a more intensive agriculture. Clan land, in turn, is shared among the members of the clan, normally in the form of reserve land. Village common land or *nagari* land is land within the village that has not been allocated to a particular clan, aside from *nagari* public land, *nagari* land is also in the form of reserve land and forest. This land is controlled by a council of elders, mostly men.

The linkage between these various types of land ownership and inheritance in case of disaster is complex and nuanced. Disaster does not directly change the inheritance of land, as owners would re-identify their former plot post disaster and enforce the ownership by re-establishing the land boundaries. It does affect those with weaker social capital and weaker socio-economic position.

4.4 The Flash Flood of 2012

A flash flood hit Padang city on July 24, 2012, at 18.30 hours. The disaster coincided with the month-long Ramadhan fast and people were breaking fast when the flood hit. Three hours of heavy rain caused the two major rivers in Padang—Kuranji and Arau—to overflow. River water collected mud and tree trunks and travelled at 40 km/hour to hit the bridges, roads, farms, plantations, houses, and livestock of Padang. As much as 146 ha of irrigated rice fields (*sawah*) in Pauh sub-district were destroyed.

In the study site, people live and farm in a narrow valley along the river. Under normal circumstances, the forest in the area would have offered protection and the river would not have been a threat. Recently, however, forest conditions have changed significantly as has the climate; rainfall no longer follows earlier patterns and water flow is no longer predictable. However, since the dangers associated with floods had not been anticipated, settlements and farmlands along the riverbank had no protection.

A flash flood is a sudden local flood, caused typically by heavy rain. According to a report produced by the city mayor's office, damages caused by the Padang flash flood included:

- Bridges, namely Koto Panjang bridge drift (resulting in a loss of IDR 7.2 billion), the suspension bridge at Batu Busuak village (loss of IDR 12.6 billion), Andalas University (Unand) Campus Bridge (loss of IDR 3.6 billion), Gunung Nago Bridge (a loss of IDR 14.4 billion), Padang Besi Bridge (a loss of IDR 1.8 billion), Sarasah Bridge (a loss of IDR 5.1 billion), and the Timbalun Bridge (a loss of IDR 3.6 billion). Total losses from damage to bridges amounted to IDR 54.42 billion.
- Road infrastructure, where damage to 16 segments resulted in a loss of IDR 30.175 billion.
- River infrastructure losses totalled IDR 135 billion from 10 rivers.
- Seven irrigation systems were heavily damaged, washed away, or had their elevation eroded, resulting in a total loss of IDR 33.5 billion.
- Three schools suffered damage to furniture, library books, computers, and fences, resulting in a loss of IDR 9.2 billion.
- Farms, plantations, and forests totalling 146 ha were damaged. Cows (22), goats (12), ducks (100), and chicken (285) were killed or washed away. One hectare of grazing land and six cattle sheds were ruined, all totalling to a loss of IDR 5.54 billion.
- In Pauh sub-district, 64 houses were heavily damaged, 48 had mild damage, 17 had minor damage, and one prayer house suffered minor damage, totalling to a loss of IDR 580 million.
- In Lubuk Kilangan sub-district, 39 houses were heavily damaged, one had mild damage, and 167 had minor damage; one office was heavily damaged, two mosques were damaged, and one prayer house had minor damage, totalling to a loss of IDR 1.44 billion.
- In Lubuk Begalung sub-district, four houses were heavily damaged, four had mild damage, and 161 had minor damage, totalling to a loss of IDR 900 million.
- In Nanggalo sub-district, one house was heavily damaged, 53 had light damage, eight mosques were severely damaged and four prayer rooms had minor damage, though the damage has not been quantified.
- In Bungus Teluk Kabung sub-district, six houses were damaged totalling a loss of IDR 300 million.
- In Padang Timur sub-district, 18 houses were mildly damaged resulting in a loss of IDR 360 million.

Total losses were estimated at IDR 263.9 billion (approximately USD 26.39 million) (minangkabaunews.com 2012).

The disaster brought to light evidence of timber felling in upper forests as logs were washed into lower-lying settlements and farmlands. Parts of farmlands became unproductive as they were filled with sand, stones, and forest debris. The floods also changed the course of the river and destroyed rice fields along the riverbank. Some of the destroyed *sawahs* were recoverable but many others were lost for good. Some irrigation facilities were damaged and the local community could not afford repairs. Some *sawahs* have now dried. Some farmers had to lose part of their harvest until sections of their *sawahs* could be restored, while others changed crops or left the villages. Most rice fields along the riverbank were washed out, eroded, or left filled with debris. This had a serious effect on land and tenure depending on the degree of damage and land ownership.

Following the floods, Agricultural officers of Padang City recorded the damage caused. Leaders of local farmers' group under the instruction of the agricultural agency, recorded the name of the farmer whose land had been damaged, the type of damage, the size of the land, and the address of the farmer. It then differentiated the agricultural damage into: eroded rice field, buried rice field, buried crop (rice), buried crop (yam), small holder plantation and horticultural land, and forestry seedling. Gender and type of land tenure and cropping arrangements were not recorded. The main purpose of the record was to assess damage. Table 4.1 presents the agricultural damage in Pauh sub-district, Padang, by *kelurahan* (smallest administrative unit in an urban area).

The record created by Padang's agricultural agency named landowners under five categories. Ancestral land under the control of the *mamak*, ancestral land under the control of *mande* (mother), ancestral land cultivated by nieces and nephews, and ancestral land under sharecroppers. However, the record did not include information on who cultivated the land.

Table 4.1 Agricultural damage caused by the July 24, 2012, flash flood in Pauh sub-district, Padang City, Indonesia by *kelurahan*

Type of damage	Administrative Unit (*kelurahan*)		
	Limau Manis	Lambung Bukit	Kapalo Koto
1. Eroded rice field (ha)	13.05	9.75	9
2. Buried rice field (ha)	16.35	–	–
3. Buried crop (rice) (ha)	–	2	3.5
4. Buried crop (yam) (ha)	–	1.25	–
5. Small holder plantation and horticultural land	Tree crops uprooted (coconut, durian, chili, long bean, tomatoes, jengkol)	–	173 trees of mixed perennial crop uprooted (coconut, clove, mangosten)
6. Forestry seedling	–	–	5000 mahogany seedlings and 5000 rubber seedlings washed away

Source Personal communication, agricultural office, Pauh sub-district, Padang, Indonesia

4.5 The Effect of the Flash Flood on Women's and Men's Land Rights in West Sumatra

Disasters and the subsequent changes wrought in land tenure affect women and men differently depending on land ownership and the damage caused. In a matrilineal society like Minangkabau, men control land in the name of kinship (*kaum*) and women own the land in cultural terms. They have complete access to the land. As mentioned earlier, the Padang agricultural agency inventoried all land and agricultural damage in the affected areas. The data included the name of the property owner; gender could be identified from the list. This list of landowners was later used by the government as the basis to provide recovery subsidies. Table 4.2 shows property damaged, by gender, as recorded by government officials.

Table 4.2 shows that more men land owners were affected by floods than women land owners. This shows that even in the matrilineal community, men are recorded as land owners. Clearly, land ownership on record differs vastly from the practice and cultural notions of land ownership. Since post-flood assistance was given to the official land owner, women lost rights to both the land as well as the assistance. We wanted to find out if the involvement of local leaders protected women's rights to land.

According to the records created by the agricultural agency, women owned one-third of the property (land) recorded. The recording was made by leaders of farmers' groups in collaboration with field extension workers. We interviewed them to understand the criteria used to determine the ownership pattern of a piece of land. We found that each group had different ways of recording land ownership.

For example, the leaders of the Harapan Jaya farmers' group considered land ownership the basic criteria when listing landowners. Land is recorded under the

Table 4.2 Type of damage as inventoried by government agency, by the sex of the owner

Type of damage	Sex of the owner			Total
	Men	Women	Group owned	
Buried paddy	4	4	0	8
Buried *sawah*	39	21	0	60
Buried yam on *sawah*	1	1	0	2
Eroded dry land	8	0	0	8
Eroded *sawah*	41	23	0	64
Livestock washed away	2	1	1	4
Uprooted tree crops	7	2	0	9
Total	102	52	1	155

Source Personal communication, agricultural office, Pauh sub-district, Padang, Indonesia

name of *mamak*, *mande*, nephew, or niece. Sharecroppers were not recorded. In Padang, sharecroppers were not recorded as victims of land damage; only landowners were. Often, it is the *mamak* who is recorded as a landowner, though he is not the owner and merely the manager of his sister's lands. Another farmer leader reported that when a person owned two plots of land, he recorded the share cropper as the owner for one property, hoping that any emergency relief provided could benefit both families. Thus different concepts of ownership are used to make official records, which then confer total control of the property to the owner.

Thus we have seen how even local participation does not ensure that the matrilineal property ownership is correctly recorded, reported and compensated. In a patriarchal system, it is men who own property, mainly the head of a household. Therefore, when people from patriarchal communities create records of landownership, it is quite likely that all property will be recorded in the name of men, not women. Thus, the shift in ownership is now confirmed in official records. Similarly, the investment of share croppers will not be compensated in the event of disasters, and since share croppers are usually landless farmers, the disaster will leave them without any resource. To recover from the loss, some people took up intensified cultivation of available land, while others were forced to find jobs outside the village. A few people took up stone and sand mining as alternative livelihoods.

4.6 Gender and Disaster Management

Mitchell (2011: ix) defines response to disaster as the "provision of emergency services and public assistance during or immediately after a disaster in order to save lives, reduce health impacts, ensure public safety and meet the basic subsistence needs of the people affected." This chapter presents the responses of the government, NGOs, and affected women and men to the flash flood in Padang and the efficacy of these responses.

The GoI through BNPB, the local government, and local people responded to the flash flood in Padang in various ways. A multi-sector approach was implemented to deal with the disaster and post-disaster rehabilitation.

Local agencies recorded all agricultural, animal, small holder plantation, and forestry damage caused by the flood. The inventory was conducted a day after the flood in cooperation with farmers' groups. Its main objective was to provide immediate data on extent of damage. Data collected included type of damage, name of victim, and amount of damage either in acreage, number of trees, livestock, or seedlings damaged or washed away. The recorded victim was usually the landowner or clan guardian (*mamak*). Emergency assistance, including food and clothes, was provided based on this list. The local government distributed food and clothes to the victims. The private sector and other institutions also followed suit. The agricultural and food crop service of Padang provided funds to restore inundated rice fields, but the most heavily damaged *sawahs* were not given any support.

Repair of irrigation canals, roads, and bridges was immediately taken up by local disaster-management agencies. Though regulations dictate that small-scale irrigation systems should be managed by water users' associations, the local government intervened to ensure that such systems were operational again. The community also took necessary action to avoid future disasters including a ban on clearing forests upstream and closure of all wood-related industry in the village. Farmers restored farmlands and continued cultivation. Some rebuilt land boundaries and put up live fences by planting trees and lemon grass. Relying on their collective memory, local people also redrew land boundaries.

By collecting data on government and community responses to a disaster during the emergency, recovery, and reconstruction stages, we discovered that land tenure is closely related to recovery from disaster. Land tenure system is defined as "sets of formal or informal rules and institutions which determine access to and control over land and natural resources" (Mitchell 2011: vii). A flash flood in a matrilineal society in West Sumatra clearly illustrated that while women land owners, women cultivators, and poor sharecroppers were the real victims, it was the men who were compensated for the loss because access was managed through the men.

Disaster management needs a gender and class sensitive participatory approach. Local people know their social relationships well, they are close to each other, they are indigenous to the place, and they have lived there for generations; they must thus be involved in disaster preparedness, management, emergency, rescue, and recovery. However, to ensure that local bias does not infiltrate the process, a clear mandate in favor of women, indigenous people and landless farmers needs to be included in the policies.

In the cycle of disaster management, action taken during the emergency stage is crucial. It determines what follows in the recovery and rehabilitation stages. Indonesian law on disaster management stipulates in article 46 that emergency reaction requires rapid and accurate assessment of the site, degree of damage, sources of disaster, emergency level, rescue and evacuation of people, and protection of all facilities and infrastructure. However, we have seen that the rapid and accurate assessment done with local participation is also prone to male bias. Therefore, disaster preparedness stage becomes critical to ensure that land rights are not alienated during and after a disaster. The usual goal is to assess damage in monetary terms to determine the scale of a disaster. However, a disaster does not affect everyone equally and the differentiated impact of disasters is mediated by gender and class.

4.7 Gender and Land Tenure in the Disaster Management Cycle

Disaster preparedness is clearly the most important part of the cycle. Land tenure in this region is not yet completely documented, and much land is held under customary law. Such rights need to be documented and more categories of land ownership are required to accurately reflect the local patterns of land holding. A cadastral survey followed by a PRA will reveal the boundaries of each piece of land; however, without special effort, women will lose their customary land rights as clan lands get registered in the name of an individual. Creating new types of property ownership such as family land and clan lands would protect the rights of women and girls in the community because individual land ownership can be sold and alienated easily, whereas clan land rights are inalienable.

In the post-disaster scenario, the emergency stage involves assessing and recording damage and conducting rescue and evacuation. During this phase, land ownership and crop ownership need to be separated to ensure that (1) the land tenure record is correct; and (2) landless farmers are compensated as sharecroppers. Therefore, compensation payable for damaged property may be recorded in favour of a people including guardians, inheritors, cultivators, or sharecroppers.

Often, disasters downgrade the quality of resources on which people rely for their livelihood. Access to and control over land constitutes a vast resource. Disasters may alter resource mapping with some farmers losing land and others gaining land, as was the case when the river changed its course or a project was taken up to widen the river. Since then, the riverbank was taken over by the people whose land bordered the river. Inclusion of share croppers and landless farmers in the emergency list would make this reallocation of land more equitable.

In the rehabilitation stage, gender and land tenure came into play again as debates rage over the rehabilitation of *sawahs*. The flash flood also changed the border of *sawahs*. New land and funds are often taken by the strongest and best-connected people if systems are not put into place to ensure that the people who have lost land will be allotted land first. In Padang, funds provided by the government to restore *sawahs* were given to those identified as owners during the emergency stage, two thirds of them were men. During the rehabilitation process, land is often required for recovery and resettlement. In Batu Busuak, a participatory approach was applied for the resettlement project. Local people were consulted and after discussion, they decided to use communal land for resettlement. A possibly contentious issue was thus amicably resolved.

4.8 Conclusion

West Sumatra province witnessed 14 flash floods in 2012 and the number is predicted to increase in future (BPBD 2012). Unfortunately, little has been done to include women in disaster preparedness and management and lesser still to ensure that cultural concepts of land ownership are maintained after disasters. Minangkabau has an egalitarian land tenure system which allowed both women and men to access land to meet their needs. The clan land was owned by women, and administered by men. With disaster and with subsequent recovery measures, the system is destabilized. Land that was owned by the clan is now being owned by individuals.

Clearly, external support from the government and NGOs has strengthened patriarchal ownership of land. The case highlights the need for a better understanding of gendered land ownership and alternate forms of land ownership so that recovery efforts do not weaken the land entitlements of women and the poor, including share croppers and tenant farmers.

References

Benda-Beckmann, Franzvon; Benda-Beckmann, Keebetvon, 2004: *Struggles Over Communal Property Rights and Law in Minangkabau, West Sumatra.* Working Paper No. 64 (Halle/Saale: Max Planck Institute for Social Anthropology).

BPBD, 2012: Various Posts of the Regional Disaster Management Agency, Sumatra are available online at: http://pusdalopspbsumbar.blogspot.com.

Kato, Tsuyoshi, 1978: "Change and Continuity in the Minangkabau Matrilineal System", in: *Indonesia*, 25 (April).

Kato, Tsuyoshi, 1982: *Matriliny and Migration: Evolving Minangkabau Traditions in Indonesia* (Ithaca: Cornel University Press).

minangkabaunews.com, 2012: *Inilah Akibat Bencana Banjir Bandang di Padang* (These were the result of the flash flood in Padang City, 24 July 2012); at: http://minangkabaunews.com/artikel-2459-inilah-akibat-bencana-banjir-bandang-di-padang.html (26 October 2013).

Mitchell, David (with the supervision of Adriana Herrera Garibay), 2011: *Assessing and Responding to Land Tenure Issues in Disaster Risk Management.* Training Manual (Rome: FAO).

Mongabay Indonesia, 2012: *Kaleidoskop Bencana Lingkungan 2012: Degradasi Hutan Melaju, Banjir Menerjang Manusia* (Environmental Disaster Kaleidoscope 2012: Drove Forest Degradation, Man Crashing Flood), Mongabay.co.id, at: www.mongabay.co.id/2012/12/28/kalesidoskop-bencana-lingkungan-2012-degradasi-hutan-melaju-banjir-menerjang-manusia/#ixzz2IaKL954C.

Naim, Mochtar, 1984: *Merantau, Pola Migrasi Suku* Minangkabau (Metrantau, Pattern of Migration Among Minangkabau People) (Jogjakarta: Gajah Mada University Press).

pusdalopspbsumbar.blogspot.com 2012: *Sumatera Barat dan Potensi Bencana 2013* (West Sumatra Province and Potential disaster 2013); at: http://pusdalopspbsumbar.blogspot.com/2012/12/sumatera-barat-dan-potensi-bencana-2013.html.

Stivens, Maila, 1996: *Matriliny and Modernity, Sexual Politics and Social Change in Rural Malaysia* (Sydney, NSW, Australia: Allen & Unwin).

UN Office for the Coordination of Humanitarian Affairs Regional Asia Pacific, 2011: *Indonesia: Natural Hazard Risks—National Hazard Map* Issue 1 March 2011; at: http://reliefweb.int/map/indonesia/indonesia-natural-hazard-risks-national-hazard-map-issued-01-march-2011 (28 October 2013).

West Sumatra Province Disaster Management Agency, 2012: Sumatera Barat dan Potensi Bencana 2013 (West Sumatra and potential disaster 2013); at: http://pusdalopspbsumbar.blogspot.com/2012/12/sumatera-barat-dan-potensi-bencana-2013.html (26 October 2013).

Chapter 5
Urbanization and Disaster: Loss of Women's Property Ownership in Leh, Ladakh

Bhuvaneswari Raman

Abstract The Himalayan region along the northern border of India has been identified as a high-risk zone, vulnerable to earthquakes, landslides, flashfloods, and drought. Drawing on evidence from the mountain town of Leh in the state of Jammu and Kashmir, this chapter explores how State interventions to promote development affect women, particularly with respect to their property relations. This chapter elucidates that development efforts have aggravated the region's vulnerability to natural disasters. A focus on tourism and associated urban development has affected women's land ownership as well as adversely affected the environment, which in turn has increased women's vulnerability to disasters in multiple ways. Women's property claims are not supported by State law or community practices, compounding their difficulty in disaster response and recovery. We argue for a better alignment of strategies governing urbanization and disasters to mitigate risks and improve disaster responsiveness. We also suggest the need for a shift in strategies, for disaster-risk mitigation to move beyond relief operations. Further, recognition of a heterogeneity of tenure forms that allow women, migrants and relatively weaker groups to establish legal claims on property is required.

Keywords Urbanization · Disaster · Gender · Land rights · Leh · Tourism

5.1 Introduction

Mountain towns are vulnerable to disaster risks arising from natural forces, and aggravated by human actions. The Himalayan region along the northern border of India has been identified as a high-risk zone, vulnerable to earthquakes, landslides, flashfloods, and drought. Its location and the presence of three perennial rivers

Bhuvaneswari Raman, Associate Professor, Jindal School of Government and Public Policy, Jindal Global University, Sonepat, 131001, India; Email: raman.bhuvaneswari@gmail.com.

K. Kusakabe et al. (eds.), *Gender and Land Tenure in the Context of Disaster in Asia*, SpringerBriefs in Environment, Security, Development and Peace, DOI 10.1007/978-3-319-16616-2_5

make this region strategically important for India in terms of defence and control of natural resources. This has influenced investments in large infrastructure projects including roads, bridges, and telecommunication. Currently, towns in the region are urbanizing rapidly chiefly due to tourism growth (Hewitt/Mehta 2012; Sharma 2008). Urbanization patterns including density of congregation, forms of land development, and rise in land prices are influencing changes in both property ownership and gender relations. It manifests as increased conflicts over property (Pirie 2007). Perceptions and practices relating to property claims are shaped by cultural norms that favor patrilineal claims. This chapter explores how State interventions to govern urbanization and disasters affect women, particularly their property relations. It argues for better alignment of strategies governing urbanization and disasters to mitigate risks and improve disaster responsiveness.

The manner in which human actions are organized in space, and resources such as land and infrastructure are distributed between community members is critical for disaster vulnerability and management in cities. For example, urban flooding, landslides, and vulnerability to earthquakes are linked to deficiencies in basic infrastructure such as drainage and rapid growth of settlements. An added complexity for mountain towns, especially in the Himalayan range, is the risk of flooding due to climate change. Rapid expansion of settlements (by clearing forests in mountainous areas) and inadequate basic infrastructure have contributed to flashfloods, submergence, and landslides. Thus, the governance of urbanization affects the disaster vulnerability of towns.

Despite this, governance of urbanization, particularly regulation of land use and development and disaster-risk management are enacted as discrete agendas. Regional and state governments have formulated disaster-management plans for each city and have constituted the institutional infrastructure to implement it. A national disaster management authority and similar structures at the district and local level have been set up. Substantial funds have been allocated for disaster management in the tenth five-year plan (GoI 2002). However, this focus on managing post-disaster situations needs to make way for disaster-risk mitigation and move beyond mere rescue and relief operations.

Further, disaster management and urban planning must be aligned and made responsive to gendered needs. Increased land prices and real estate development have altered traditional community-based land use governance and property claims. Such changes adversely affect the territorial claims of women with relatively weaker bargaining power, as well as those of different groups in normal times, more so in post-disaster situations. Gender insensitivity in urbanization and disaster-management policies, together with gender inequity in the property laws of the State and community aggravate women's position in responding to disasters.

The primary data for this chapter was collected during a long-term ethnographic study, which was then followed with detailed interviews conducted in the second half of 2013, with land owners (women and men) as well as tenants and government officials to collect qualitative data on the impact of disasters on women's land rights. This chapter elaborates on the above argument drawing on evidence from the mountain town of Leh in the state of Jammu and Kashmir. Spatial and social ecology of the town, disaster risks in the area, land tenure and gender relations, and State interventions in disasters and land development and their implications for property and gender relations are also discussed.

5.2 Ecology and Disaster Risks in Leh

Leh town is purposively chosen as the study area for the following reasons: First, mountain towns in India are experiencing rapid urbanization but research on them is limited. Second, the use and control over land resources is shaped by both modern and customary laws in Leh. Third, the Himalayan region where Leh is located has been identified at high risk of earthquake and floods, following melting of glaciers due to climate change. Fourth, Leh was built by Buddhists, and continues to be dominated by them. As such women have strong rights to land unlike the rest of India.

Leh's history can be traced back to the fifteenth century. It was located along the silk trade route connecting Central Afghanistan, India, and Tibet (Rizvi 1999). Due to its border dispute with Pakistan and China, the Indian government maintains a large military base in Leh (Goodall 2004). Leh is a fast urbanizing town in Ladakh district; it recorded an average population annual growth rate of 5.2 % between 1981 and 2001, nearly double the national average rate (2.95 %) of urbanization in India during the same period.

Leh is highly vulnerable to earthquakes and flooding (on account of rivers changing course frequently). Rapid urbanization and the development agenda promoted by the Indian government have heightened disaster-related risks. Leh is strategically located on the frontier zone bordering China, Tibet, and Pakistan. The Indian government's investment in infrastructure, particularly connectivity (roads and telecommunications), in Leh is driven by security concerns. The perennial river Indus flows through Leh, and contestations around sharing of river water with neighbouring countries influenced the Indian government's decision to invest in dam construction along the course of the Indus. Tourism is promoted as the main economic development strategy in Leh.

5.3 Spatial and Social Ecology of Leh

Leh town is the capital of Leh district in Ladakh, situated in the Indian state of Jammu and Kashmir. Ladakh is located at a higher altitude; since it is in the lee of the Himalayas it receives scanty rainfall. Its landscape is made of steep slopes and minimal forest cover, with very little arable land. The 'cold desert on the mountains', Ladakh experiences extreme cold and dry weather, with high radiation, strong winds, and low humidity. Land available for cultivation and settlement is limited. Residential settlements have developed in the valleys along the Indus river and its tributaries, Shyok/Nubra and Zangskar. Ladakh also has a number of lakes and two important passes—Khardung La and Zojila—that connect Ladakh with the rest of the country. Heavy snowfall along the two passes in winter makes Ladakh inaccessible for nearly 6 months in a year.

Leh, comprising six blocks, is the second largest district in the country. Leh district houses nearly three-fourths of Ladakh's urban population. In 2001, the district's population was 147,000 of which nearly 20,000 resided in Leh town. The district, particularly Leh town and its surrounding areas, has been urbanizing rapidly. The 2001 census registered a 23 % increase in urban population, excluding the floating population in Leh town. In 2011, the increase was just about 13 %. Ladakh is predominantly a rural society; nearly 68 % of its population resides in rural areas (Census 2011). Historically, local livelihoods involved subsistence agriculture and animal husbandry (mostly goats, sheep, cows, yak, and dzomo). The chief occupations include cultivation (38 %); agricultural labor (4 %); household industry (1 %); and government and services, chiefly tourism (57 %).

Ladakh's ancient inhabitants were Dards, an Indo-Aryan race. Over time, immigrants from Tibet, Skardo, and adjacent areas like Purang and Guge settled in Ladakh. According to the 2001 census, the population of Leh is made up of Tibetan Buddhists (around 77 %), Shia Muslims (around 14 %), and Christians and Hindus (9 %) (Census 2001). Although Ladakhi society is considered egalitarian in terms of gender relations, the child sex ratio for 0–6 years was around 946 females to 1000 males in Leh district, and the adult sex ratio registered 826 females to 1000 males (Census 2001). The disparity in adult sex ratio is attributed to seasonal migrant male workers; the town's population fluctuates and reportedly doubles during summer, which is the peak tourist season (Alexander/Catanese 2007). Besides nomadic communities from surrounding districts who congregate in Leh during certain months, the town attracts tourists and traders from the Kashmir valley for 6 months in a year. The traders, predominantly Sunni Muslims, frequent this town in summer for trade in woollen products and carpets. During the peak summer season in 2006, an estimated 50,000 people visited the town; their numbers keep increasing over the years (Alexander/Catanese 2007).

Leh town is the administrative capital and a military base. It was built as a fortified city in the early fifteenth century by a Ladakhi king and declared the capital of Ladakh kingdom in the seventeenth century. The royal residence and the fort atop the hill formed the focal point of the town, followed by the residences of the courtiers and the elite. The houses of the commoners were at the foothill (WMF 2008). The old town has around 200 houses of mud, stone, and timber connected by a maze of alleyways. Thus, the spatial organization also reflected the old town's social hierarchy. Over time, large landowners moved away from the hill to their fields on the plains.

Lack of investment in public infrastructure and private properties led to the dereliction of the old town. Wealthier property owners moved out of the old town and converted their agricultural land into residences and/or hotels and lodges. These economies attracted migrant labor, particularly from Bihar. Many stay within their place of work. The middle income households, predominantly Muslims, occupied vacant land in the foothill and developed their residential colonies. With the movement of the wealthier population and the middle income households to the foothill and plains, the middle and lower income households were trapped on the mountain. A survey undertaken for a project to conserve the old town reported that the average monthly household income in 2004 varied between 2,000 INR (40 USD) to 6,000 INR (120 USD) (Alexander/Catanese 2007). Currently, old town properties are occupied by Muslims and Buddhists who work as skilled labour in construction, hotels, workshops, or in lower-level government jobs.

At the time of our field research, properties on the higher slopes of the hill, held by some of the Buddhist elites were dilapidated and inaccessible. In the event of a landslide or flood, these dilapidated structures could damage residences occupied by low-income households on the lower slopes. The survey of the old town (WMF 2008) identified several relevant issues at the intersection of gender, disaster and land tenure, including poor living conditions due to lack of basic infrastructure and inability to invest in repairs; decline of old town; government inactivity; landless people trapped in poverty; loss of cultural identity due to migration and rapid economic changes; and property-related tensions between different religious communities.[1]

5.4 Disaster Risks in Mountain Towns and Impact of Development

Policy responses to disasters tend to focus on crisis management. However, analysis of the causes of disasters in Indian cities highlights that many natural or man-made disasters stem from inadequate or inappropriate planning choices and failure to invest in basic infrastructure facilities such as storm water drainage and solid waste management. These result in flooding and droughts. Unregulated land development, specifically on earthquake-prone terrain, leads to quakes, as evidenced by the Bhuj earthquake and the Uttarakhand floods. Table 5.1 summarizes the disaster threats faced by Leh town.

Leh town was one of the worst affected by the floods in 2011, particularly the old city. Low-income households who predominantly occupy residential structures in the old city were among the worst affected in the district. There are varying estimates of damage to life on account of flash floods. While official reports counted the death toll at 286 and those with grievous and minor injuries at 206, NGO estimates put the death toll at 234 while 824 were reported to be seriously injured (Oza/Prajapathy 2012). The flashfloods that carried muddy water from the hills along three different routes submerged houses and washed away everything, leaving silt and boulders. The floods also disrupted communication infrastructure. The intensity of the disaster

[1]Although urbanization and lack of urban planning is a serious issue in this area, the focus of this chapter is on how disaster affects women's land ownership.

in the old town was aggravated by the dilapidated condition of houses on the slope and lack of investments in basic infrastructure.

As can be inferred from Table 5.1, unregulated development of land and inadequate infrastructure has accentuated the disaster risks in mountain towns. Urban floods have been recognized as a key threat in the Indian government's disaster-management plan since 2005. Flooding is caused by the State's failure to invest in basic infrastructure (Gupta/Nair 2011). Melting glaciers and incessant rains over 2 days caused flashfloods in Ladakh in August 2011. The floods resulted in loss of property, livestock, and human lives. Nearly half the casualties were local residents, aged 25–50 years. Though gender disaggregated data is not available, it is estimated that more than a third of the affected persons were women aged 10–50 years (Oza/Prajapathy 2012; Gupta et al. 2012). The number of women who were affected is lower than that of men because of the large number of male migrants and the low sex ratio. However, even though the actual number of women casualties was low, the impact they experienced was harsher than that of men, as we will discuss below. Table 5.2 below summarizes the extent of damage.

Leh town has been identified as a high-risk earthquake zone. It is situated on a strike-slip fault line. Unregulated clearing of mountain slopes, conversion of agricultural land for homes and tourism infrastructure, and rampant construction of houses on canal beds and dry canals could accentuate the disaster risks (including earthquakes, floods, landslides, and snowslides) in Leh town.

Tourism is the main development strategy and Leh's tourism history dates back to the 1980s. The influx of tourists has placed a heavy strain on environmental health. One challenge is the management of solid and liquid waste. Basic infrastructure provisions—sewage, garbage disposal, waste-water purification—have not kept pace with the rapid increase in consumption. Inevitably, liquid waste goes back into the ground, while solid waste is disposed off on the streets. Tourism

Table 5.1 Disaster risks and driving factors

Disaster risk	Definition	Factors that exacerbate effects of disasters
Cloudburst and flashflood	Extreme weather event in which very heavy rainfall occurs over a highly localized area in a very short time	Encroachment of *nullahs* (drains) and tank beds; occupation of low-lying areas; melting of glaciers (climate change-induced risk)
Avalanche, landslide, snow, and road blockade	Sliding of snow from upper slope due to lack of gravity. The two passes leading to Leh have been identified as high-risk zones. Ladakh's geology is characterized by loose soil that has sedimented over several years of erosion	Deforestation (clearing of slopes for residential development); inadequate drainage; and density of development
Earthquake	Listed as zone IV or high-risk earthquake zone	Land use—density, materials, and infrastructure provision

Source GoJK (2011)

Table 5.2 Impact of the flashflood of August 2011 in Ladakh district

Description	Numbers
Number of villages affected in Leh District	71
Population affected	9,000
Damage to agricultural land and crops	
Total land area affected	687 ha
Total cropped area affected	660 ha
Area where crop damage was greater than 50 %	Leh district
Estimated value of damage to public property due to floods	Rs 1.33 billion (USD22 million)
Houses damaged due to floods	
Fully damaged *kutcha* (temporary materials) structures	664
Partly damaged houses	783
Huts damaged	45
Total number of houses damaged	1,447
Total number of commercial structures	458
Livestock lost	402

Source Oza/Prajapathy (2012); Gupta et al. (2012); LDC (2011)

contributes to nearly 50 % of the district's GDP (Pelliciardi 2010). In 2010, Ladakh district had around 122 hotels and 274 guest houses, mostly concentrated in the town of Leh (6400 beds in 250 hotels and guest houses); about 170 tourist agencies are registered as members of ALTOA (All Ladakh Tour Operators Association) to service its more than 70,000 annual visitors (Pelliciardi 2010).

Another concern is the uncontrolled drilling of bore wells to augment water supply in Leh town. Many business establishments have dug deep bore wells that are rapidly depleting groundwater levels. Guest houses have installed water-guzzling flush toilets. Most Ladakhi households, especially in the old town, use traditional soak pits, the waste from which is collected in compost holes to be used as fertilizer. Ironically, however, even businesses with flush toilets are forced to maintain traditional soak pits, as the water in the pipes freezes during Leh winters. The lack of adequate infrastructure makes Leh more vulnerable to flashfloods and earthquakes.

New houses have been constructed using contemporary design and materials. The move away from traditional design and materials (characterized by light mud roofs and walls) to concrete structures has decreased the community's resilience in responding households to climate change related risks. As argued by Singh (2011), Ladakh's traditional settlement design and construction techniques aimed at minimizing floods. For example, the Mani wall, designed as a sacred wall, performed a functional role; the mud walls and roofs were designed to withstand 6 months of cold.

In Leh, Ladakh, interviews revealed that Muslim women and women migrants from poor families faced additional difficulties as rehabilitation was very slow. Women's recovery of land is even slower than men's because community leaders who distribute land are all men, and cultural norms limit women's interaction with male leaders. When inefficient land recovery processes are compounded by state/community patriarchy, women's vulnerability increases and their control over land reduces further.

5.5 Land Tenure, Gender and Property Relations

Land in Leh town is controlled by the State and the civilian population. Due to the strategic location of the town, the military controls large tracts of land. A substantial proportion of territory in Ladakh, particularly Leh town is controlled by various State agencies, particularly the Ministries of Defense and Forests. Several key military installations are located in Ladakh, which serves as the strategic frontier for India because of its proximity to Pakistan and the political instability in Kashmir. The Ministry of Defence's research lab is in Leh town and, the Border Security Force has a large presence (Planning Commission 2003).

Among civilian land owners, institutions, particularly Buddhist monasteries and the Buddhist elite, control most of the land in Leh. Individual land parcels are predominantly owned by the Buddhists. Monasteries and large landowners lease land to several small farmers, including both itinerant local populations as well as migrants.

Both customary practices and State laws are underpinned by patriarchal values of property. Customary practices promote inheritance of land and property along patrilineal lines in Leh. It is useful here to understand the status of women in Leh. Women form the backbone of the economy as they are responsible for the production of food to meet household needs. They have the freedom to select partners. As bride price is important among the tribes of this region, a groom unable to pay has to either work for the girl's family for a period stipulated by the village council or arrange a bride for the girl's brother or cousin (Bhasin 1999a, b). Similarly, pregnancy before marriage is accepted and does not necessarily affect the girl's chances of finding another partner.

The division of labor, however, is similar to other patriarchal societies. Girls are expected to participate in all types of work at home and in agricultural activities along with their mothers. They act as pseudo-parents and look after their siblings. If they have spare time and parental permission, they can go to school. However, the situation varies across different income groups. As Bhasin (1999a, b) notes, women's supremacy is restricted to the household and does not extend to public realms. Women's economic participation does not translate into corresponding community authority.

While women are free to choose their partners or exercise sexual freedom, property relations are defined by patrilineal traditions in Leh. In Buddhist families, property commonly passes onto the eldest son while daughters inherit ornaments (Bhasin 1999a, b). In Ladakh, daughters inherit property only in the absence of a son. At the time of their marriage, daughters may enter into a type of negotiation known as *magpa* which guarantees them sole right to property. The husband has no right over such property and moves into his wife's residence after marriage. With the introduction of laws for abolishing large land estates, the terms for sharing property between siblings have changed.

Customary laws are enacted by community institutions organized on different territorial scales in Ladakh—the neighbourhood or village, a group of villages, or a city. In the villages around Leh, society is organized around several actors, including the *gopa* (village head), *onpa* (tax collector), *chupon* (official in charge of water distribution), *nyerpa* (in charge of festivals and religious activities), and the *gompa* (monastery) or mosque (Pirie 2007). The *gopa* liaises with the State over issues of land and infrastructure allocation and mediates the allocation of land between members of a community and between different communities (Pirie 2007). By virtue of their role as tax collectors, *onpas* maintain records of land claims and land-use practices. In Leh town, Islamic and Buddhist religious communities have their own representatives (committees)[2] and a *gopa* represents the entire town. Several neighbourhood leaders engage with the city-level *gopa* and the State independently.

Customary laws govern perceptions and arrangements relating to the use of and claims to land held individually and communally. Claims are ascertained through daily practices. Often, customary laws are transmitted as oral traditions. The norms are shaped by religious and/or tribal values and practices (Pirie 2007; Goodall 2007; Choudhury 2009). Buddhist and Islamic values and their institutions influence customary laws and their practice in daily life in Leh. With rising land prices, conflicts around land claims at the intra-household level and between communities (over common land) are increasing, especially in urban centres like Leh (Pirie 2007). The implications of this real estate dynamics on property, religious control and the interaction of customary practices with modern legal practice needs to be researched further.

The dominant trend is to rely on community institutions rather than State institutions for dispute resolution. As Pirie (2007) notes, Ladakhis prefer to resolve conflicts through community institutions and opt for the State's modern legal system only as a last resort. This practice is common even among the educated class. Customary practices built through long period of cooperation between itinerant and settled communities influence the manner in which village commons are

[2]A Muslim committee headed by a *gopa* exists at the village level, playing a role similar to Buddhist *gopas*.

used. Such practices are mediated by extensive kinship ties between Ladakhis in Leh town and across different villages.

Until 2000–01, Buddhist women frequently married Muslim men in Leh. But since 2007, Buddhist monasteries have been discouraging inter-religious marriages. The pressure on Muslim men to express religious loyalty resulted in intra-household conflicts between couples from different religious backgrounds. This extended to property relations. As mentioned earlier, when a Buddhist woman married a Shia Muslim man, the former held land and any structures constructed on it. Women in such inter-religious marriages were under greater pressure to express their loyalty to their husband's faith and this conflict at times extended to clarifying property relations. Thus, women's participation in the economy or control over property provided them with limited autonomy in the private sphere and did not translate into freedom in the public realm.

The State's property laws for promoting individual land ownership and assuming males as head of household compounds the gender inequity (Fernandes/Barbora 2002; Marak 1997; Mann 1987). Further the condition for availing government welfare schemes is premised on the assumption of individual ownership of property and male-headed families further reinforce the trend of securing individual titles in the name of men.

The conversion of community-owned land to commercial development has serious consequences for the claims of poor groups who often rely on customary practices or use-based rights. In Leh, the shrinking of commons directly affects the livelihoods of pastoral communities. The move towards private individual tenure also has serious implications for women from ethnic groups whose customary claims to land are based on use and hinge on male inheritance. In such places, women's claim to land is fragile to begin with, as they only have usufruct rights, not *defacto* claims to land they use.

The local population in Leh is predominantly Buddhist and this culture gave women some resources like inheritance of land and house, though men controlled public spaces where decisions related to community land use were made. Thus, women had some land rights, though these rights were mediated by men in public spaces. Into this situation came migrants from patriarchal communities as well as the State, with several development efforts including modern land management systems and support for tourism. Migrants, who were mostly men, took control over land, and subsequently, public spaces where land-related decisions are made. The State's agenda of modernization of land ownership as well as tourism development were implemented with a patriarchal bias, and so land certificates were issued in the name of men. Due to this, women's precarious hold on land was weakened further.

5.6 Conclusion

Traditional norms supported women's claims to land under certain circumstances, however, modernization has weakened even these few rights and women's land rights are now in a precarious situation. Urban planning did not consider gendered land rights and led to the consolidation of land in the hands of men. Such skewed ownership of land towards men increased women's vulnerability to disaster in a town that is ill-prepared for disasters in the first place. There is a need to rethink issues of land tenure and the land entitlement of women and other disadvantaged groups.

This chapter makes a case for broadening the framework of disaster management to include disaster risk reduction. India has established an elaborate institutional infrastructure to address disasters. Disaster management plans have been prepared at the State and district levels. Disaster-management centres have been set up at the national and district levels. With regard to State interventions, Ladakh region has disaster-management plans in place. These emphasize dealing with the aftermath of a crisis. Disaster preparedness is thus conceptualized within the narrow framework of 'crisis management' and different departments coordinate to ensure smooth relief operations. Risks include uncontrolled real estate development and unsustainable use of natural resources in areas identified as high-risk earthquake zones. Urban planning and development policies have still not been effectively aligned with disaster management.

Disasters can frequently erase both individual and collective land claims and have serious consequences for livelihoods. The assumption in land policies is that land is held individually and as a freehold. Titles and possession certificates are issued to authorize individual user rights. By contrast, land tenure relations of pastoral communities are characterized by fluid and temporary terms of use in different places. In the event of a disaster, the land-use rights of women, small farmers and itinerant pastoral communities are getting eroded. Disaster rehabilitation strategies do not take into account their needs.

Exploring the disconnect between governance of growth, urbanization, and disaster risks, we find that strategies for dealing with urban disaster are focused on post-disaster situations neglecting everyday hazards, which paradoxically accumulates the risks of major catastrophe. As important as it is to address post-disaster situation, it is equally vital to re-examine the institutional and legal framework for urban planning to mitigate disaster risks in mountain towns. The singular focus on tourism neglecting any other socio-economic concerns have increased Leh's vulnerability to disaster. Because of the different entitlement to land and other resources as well as the patriarchal practices in the household and community, women are more vulnerable than men. There is an urgent need for urban planning to take into consideration both disaster risk management and gender equality to build a more resilient town.

References

Alexander, André; Catanese, Andreas, 2007: "Leh Old Town Conservation Project—Ladakh, Indian Himalayas", in: *E-Conservation*, 1 (October).

Bhasin, Veena, 1999a: "Leh—an endangered city?", in: *Anthropologist*, 1,1: 1–17.

Bhasin, Veena, 1999b: *Tribals of Ladakh: Ecology, Human Settlements and Health* (New Delhi: Kamla-Raj Enterprises).

Bhasin, Veena, 2004: "Ecology and Status of Women among Tribals of India", in: *Journal of Human Ecology*, 15,4: 237–249.

Census, 2001: *Census of India 2001* (New Delhi: Government of India) (5 June 2014).

Census, 2011: *Census of India 2011: Leh District* (New Delhi: Government of India); at: http://www.census2011.co.in/census/district/621-leh.html (7 July 2014).

Choudhury, Maitreyee, 2009: "Land and Forest in the Eastern Himalayas: A Critique on Agriculture and Agroforestry in Arunachal Pradesh", in: *Dialogue*, 10,4 (April–June).

Fernandes, Walter; Barbora, Sanjay, 2002: *Modernisation and Women's Status in North Eastern India: A Comparative Study of Six Tribes* (Guwahati: North Eastern Social Research Centre).

GoI, 2002: *Tenth Five Year Plan 2002–07 of Planning Commission* (New Delhi: Government of India).

GoJK, 2011: *Jammu and Kashmir State Disaster Management Policy* (Draft) (Srinagar: Government of Jammu and Kashmir).

Goodall, Sarah, 2004: "Rural-to-urban Migration and Urbanization in Leh, Ladakh. A Case Study of Three Nomadic Pastoral Communities". in: *Mountain Research and Development*, 24,3 (August): 220–227.

Goodall, Sarah, 2007: *From Plateau Pastures to Urban Fringe: Sedentarisation of Nomadic Pastoralists in Ladakh* (PhD dissertation, University of Adelaide: Department of Geography and Environmental Studies, Adelaide).

Gupta, A.K.; Nair, S.S., 2011: "Urban floods in Bangalore and Chennai: Risk Management Challenges and Lessons for Sustainable Urban Ecology", in: *Current Science,* 100,11: 10.

Gupta, Preeti; Khanna, A.; Majumdar, S., 2012: "Disaster Management in Flash Floods in Leh (Ladakh): A Case Study", in: *Indian Journal of Community Medicine*, 37,3 (July–September): 185–190.

Hewitt, Kenneth; Mehta, Manjari, 2012: "Rethinking Risk and Disasters in Mountain Areas", in: *Journal of Alpine Research,* 100,1: 2–10.

LDC, 2011: *Ladakh District Disaster Management Plan 2011* (Leh: District Administration).

Mann, Michael (Ed.), 1987: *Macmillan Student Encyclopaedia of Sociology*. (London, Basingstoke: Macmillan).

Marak, Kumie R., 1997: *Tradition and Modernity in a Matrilineal Tribal Society* (New Delhi: Inter- India Publications).

Oza, Sanchit; Prajapathi, Kalpesh, 2012: *Long Term Recovery Efforts in Cloudburst Affected Urban Area of Leh* (Ahmedabad: All India Disaster Mitigation Institute).

Pelliciardi, Vladimiro, 2010: "Tourism Traffic Volumes in Leh District: An Overview", in: *Ladakh Studies*, 26: 14–23.

Pirie, Fernanda, 2007: *Peace and Conflict in Ladakh: The Construction of a Fragile Web of Order.* Pirie Tibetan Buddhist Studies Series (Leiden: Koninklije Pirie).

Planning Commission, 2003: *Jammu & Kashmir State Development Report* (New Delhi: Government of India).

Rizvi, Janet, 1999: *Ladakh: Crossroads of High Asia* (New Delhi: Oxford University Press).

Sharma, Manish, 2008: "Rethinking of Urbanisation in Arunachal Pradesh", in: *Indian Journal of Regional Science*, 40,2: 142–150.

Singh, Harjit, 2011: "Leh Tragedy of August 2010—Some Lessons for Future Development and Probable Implications of Climate Change". Paper presented at the 15th International Association of Ladakh Studies (IALS) Conference, on 19th August 2011 at Leh, Ladakh. Abstract at: http://ladakhstudies.org/resources/Conferences/Leh-2011/Leh-abstracts.pdf.

WMF, 2008: *Leh Old Town/Leh Palace: Field Project Survey 2008 Watch* (New York: World Monument Fund); at: http://www.wmf.org/project/leh-old-town-leh-palace (20 June 2014).

Other Literature

"District Profile Leh", Official Website of District Administration, Leh; at: http://leh.nic.in/pages/leh.pdf (11 December 2013 and 12 June 2014).

Chapter 6
A Coir Mill of Their Own: Women's Agency in Post-tsunami Sri Lanka

Ramanie Jayatilaka

Abstract The 2004 tsunami devastated large parts of the coast in Sri Lanka causing loss of life and livelihoods and displacement of families and communities. Recovering lost livelihoods was a major challenge for the State in the post-recovery process. The case study presented in this chapter describes the experiences of a local women's organization that faced the challenge of restoring the livelihoods of its members, specifically coir workers. After the tsunami, the fisherfolk became the focus of aid organizations and coir workers were largely neglected. As a result, the men who dominated the coir industry were not interested in reviving it. With the support of local NGOs and CBOs, women coir workers entered this space to establish their own small-scale coir fibre-processing centre and broke the monopoly of private individuals who sold coir fibre at high prices. However, running the coir fibre mill, providing coir fibre to women for production, and negotiating a good price for the coir-based products are new challenges for the women as well as women's organizations. Disasters can offer women some opportunities to exercise their agency and challenge oppressive social norms and barriers.

Keywords Tsunami · Livelihoods · Coir work · Resilience · Collective strength

Abbreviations

CBO	Community based organizations
CNO	Centre for National Operations
DS	Divisional Secretary
GN	Grama Niladari
NGOs	Non-governmental organizations
TAFOR	Task Force on Relief
TAFREN	Task Force for Rebuilding the Nation
TAFRER	Task Force for Rescue and Relief

Ramanie Jayatilaka, Department of Sociology, University of Colombo, Sri Lanka; Email: ramaniej@yahoo.com.

K. Kusakabe et al. (eds.), *Gender and Land Tenure in the Context of Disaster in Asia*, SpringerBriefs in Environment, Security, Development and Peace, DOI 10.1007/978-3-319-16616-2_6

6.1 Introduction

The word 'tsunami' was introduced into the Sri Lankan vocabulary after 26 December 2004. Sri Lanka had not experienced such a disaster earlier although it had faced other natural disasters including floods, droughts, cyclones, and landslides. The magnitude of the tsunami shocked not only the people directly affected but also others. In Sri Lanka, about 35,322 lives were lost and 516,150 people displaced (UNDP 2005); another 21,000 Sri Lankans suffered injuries due to the tsunami. In addition, massive damage was inflicted on houses, personal assets, public property, railways, bridges, communication networks, hospitals, schools, and other infrastructure. Seventy per cent of the coastline was damaged. Numerous livelihoods, especially in the fishing industry, were severely damaged.

Disasters are inherently social phenomena and have their foundations in social systems or structures including gender, caste, class, and ethnicity (de Mel/Ruwanpura 2006; de Mel 2007; Mubarak-Perera 2013). Fothergill (1996), quoting from Quarantelli (1994), argues that although disaster researchers accept the importance of social structure, many have failed to consider the larger social context to understand disasters. Gender as a critical dimension of social structure has been neglected in disaster studies. However, recently, attempts have been made to understand women's experiences and their perspectives on vulnerabilities, risks, and post-recovery processes.

Certain groups are more exposed to disasters, more at risk, and find the recovery process more hazardous and difficult because of the economic, political, social, and cultural environments in which disasters take place. Women are more vulnerable to disasters due to discriminatory social norms and practices that restrict their mobility, access to information, and skills such as swimming, climbing, and running. The gendered care role of women further means they prioritize the care of children, elderly, and sick over their own safety (Hyndman 2009).

Gender insensitivity was apparent early in post-tsunami recovery and reconstruction activities (Ruwanpura 2008, 2009; among others). Hardly any attempt was made to collect sex-disaggregated data on fatalities, disappearances, and displacements although the Ministry of Women's Empowerment and Social Welfare had the overall responsibility for disaster management. None of the agencies entrusted with humanitarian relief—the *Centre for National Operations* (CNO) and the *Task Force for Rescue and Relief* (TAFRER) and later their replacements, the *Task Force on Relief* (TAFOR) and *the Task Force for Rebuilding the Nation* (TAFREN)—collected gender-sensitive data nor was service delivery guided by any gender perspective (de Mel et al. 2009). However, later studies undertaken by scholars commissioned by the government, multilateral agencies, or NGOs and individual researchers have been useful in understanding gender aspects.

They highlighted higher incidences of deaths, injuries, and displacements among women and girls due to the tsunami (Oxfam 2006; Emmanuel 2009; Silva 2010). "There is no accurate information about the age and gender distribution of deaths caused by tsunami, but it is generally understood that women, children and the elderly were the greatest casualties. According to one estimate, more than two-thirds of the victims were women or children under the age of 6 years" (Silva 2010: 161). Growing literature also illustrates that women coped effectively with the tsunami, thus highlighting women's agency in challenging oppressive social norms and barriers (Emmanuel 2009).

The objective of this chapter is to explore how women from low-income groups collectively faced the challenge of recovering from the loss of livelihoods (coir work) after the tsunami. It also examines their response to the challenges of sustaining the coir industry amid unfavourable market forces and social attitudes about women's mobility. The case study highlights how macro-economic forces and social disapproval of women's mobility now pose a threat to their work although these women had the capacity and strength to recover from the complete destruction of a traditional livelihood.

The study followed an interpretative research approach; the case study method was used to collect information. Qualitative data was collected through in-depth interviews with women engaged in coir work.[1] Similar interviews were conducted with key office bearers of the women's organization. Focus group discussions were conducted with women engaged in coir work and members of the women's organization. In addition, statistical information was collected from a number of sources. Observations also provided a good opportunity to understand women's engagement in coir work in the natural setting. The many visits made by the researcher to the south coast soon after the tsunami (2005) to collect people's voices on tsunami relief, reconstruction, and rehabilitation also proved helpful in understanding the experiences and opinions of women and men severely affected by the tsunami. Perusal of documents related to the organization helped in understanding its structure, objectives, and activities.

6.2 The Setting

The tsunami struck 13 of the 25 districts in Sri Lanka (UNDP 2005). Galle is one of the three districts in southern Sri Lanka that were severely affected by the tsunami. Galle has 18 *Divisional Secretary* (DS) divisions and 895 *Grama Niladari* (GN) divisions. This study is located in three GN divisions, namely Peraliya North, Peraliya South, and Thelwatta in the Hikkaduwa DS division, the most severely affected DS division in Galle. The train disaster that resulted from the tsunami and killed several people took place in Peraliya. Galle's recorded 1,559 deaths and 226 disappearances in the tsunami. The number of women outnumbered that of men in both the death and disappearance, highlighting their vulnerability to disasters.

[1]This research into the coir industry in south Sri Lanka began with the researcher's association with a national NGO that assisted local NGOs through district consortiums to recover from the tsunami and implement other developmental work.

Table 6.1 Number of damaged housing units, by ownership, in the GN Divisions of Peraliya North, Peraliya South, and Thelwatta in the Hikkaduwa DS division in Galle district (February 2005)

GN division	Total number of damaged housing units	Ownership of land				
		Belonging to the family	Rent/lease	Coastal reservation	Other	Not reported
Peraliya north	175	150	0	6	18	1
Peraliya south	120	113	1	3	3	0
Thelwatta	303	286	2	1	11	3

Source Department of Census and Statistics (2005)

Hikkaduwa DS division accounted for the highest number of deaths and disappearances in Galle district at 740 and 115, respectively (Department of Census and Statistics 2005).

The tsunami also destroyed houses and livelihoods along the coastal belt. Following the loss of loved ones, the destruction of homes and livelihoods posed a major challenge for survivors struggling to return to normalcy. Damaged houses were categorized by authorities as partially or fully damaged. Table 6.1 lists the number of damaged houses and their ownership type for Peraliya North, Peraliya South, and Thelwatta GN divisions; however a gender-based breakdown of land/house ownership is not available. Qualitative data collected for this study revealed houses and lands were generally in the names of men, mostly husbands. However, ownership seemed to have transferred to a few women after the death of their husbands. Sri Lanka has a General Law and a number of Personal Laws on land-ownership applicable in certain areas or among certain ethnic groups. In the south, women are generally governed by the General Law, which is egalitarian and gives equal inheritance rights to spouses, children, or siblings (Goonesekera 2006, 2009). However, women are deprived of their lawful rights to land and property by family members. The denial is largely in favor of male family members. Further, such deprivation of women's land rights is justified by both women and men (Pinto-Jayawardena/de Almedia Gunarathne 2010).

6.3 Destruction of Livelihoods

The tsunami destroyed not only houses but also many livelihoods. The most important economic activity for men in the area was fishing, which was severely affected by the tsunami. Women mostly undertook coir work as a home-based economic activity. They also engaged in sewing, making different eatables for sale, and petty trade. Table 6.2 lists the economic activities of the affected people in the Galle district.

Table 6.2 clearly shows that while fishing was the main economic activity for men, coir was women's main livelihood in the Galle district. However, the coir

Table 6.2 Main occupation of people living in damaged housing units in the affected DS divisions in Galle district

Main occupation	Men (%)	Women (%)	Total	
			No.	%
Agriculture/livestock	2.5	1.9	41	2.3
Fishery (for sale)	40.3	5.2	528	30.0
Other fishery related industries	4.6	2.7	71	4.0
Coir industry	4.7	35.3	242	13.7
Lime industry	4.7	–	59	3.3
Other manufacturing industries	13.1	22.2	278	15.8
Trade	13.8	18.7	268	15.2
Hotels/restaurants	4.3	4.8	78	4.4
Tourist hotels	0.6	1.0	13	0.7
Restaurants/guest houses/kabanas	0.4	0.6	8	0.5
Government service	0.1	0.2	2	0.1
Educational service	–	1.2	6	0.4
Health service	0.4	0.2	6	0.4
Other services	8.7	5.2	135	7.7
Other	0.6	0.6	11	0.6
Not stated	1.2	0.2	16	0.9
Total and percentage	100.0	100.0	–	100.0

Note n = 1243 for men; n = 519 for women
Source Department of Census and Statistics (2005)

industry was completely destroyed by the tsunami when coir pits, mills, fibre, and machines were all washed away. Women also lost other home-based economic activities such as sewing and making eatables for sale since their equipment was washed away. The loss of their livelihoods left women extremely vulnerable. The department of census and statistics reported that before the tsunami, 59, 30 and 133 people were respectively engaged in the coir industry in Peraliya North, Peraliya South and Thelwatta GN divisions of Galle district. As of February 2005, nobody was engaged in the coir industry (Department of Census and Statistics 2005).

6.4 Coir Industry and Its Characteristics

Sri Lanka produces two types of coir fibre: the traditional white and the modernized and mechanized brown. White fibre is labor intensive and is produced mainly by women in the western and southern provinces. Brown fibre is export-oriented and produced in the north-western province. Sri Lanka produces four main categories of coir fibre—bristle, mat, mixed, and mattress. They are sold either as raw materials in the international market or as brooms, brushes, twine, matting, rubberized coir mattresses, and upholstery (Oxfam 2006). The coir industry has its class and gender dimensions.

More than 75 % of the workers engaged in the coir industry in the Southern Province are females, particularly among self-employed and family workers. Both genders work in coir mills and coir factories, where wage differentials between men and women still persist despite the apparent similarity of the work they perform (Oxfam 2006: 4).

Women from low-income groups make the coir ropes, mats, and other items, while the raw material (coir fibre) is owned by *mudalalis*[2] who are men from a higher economic class. *Mudalalis* own private coir mills (coir pits prior to tsunami) and coconut groves. Some also own centres that produce desiccated coconut for export markets. They also buy coir ropes from producers and sell them to the central market for a profit. Thus, the real producers (women) have no ownership (Risseeuw 1980; Peiris 1997; Kumara et al. 2011).

In coastal areas, agricultural activities are rare and coir work is a major source of income. The extreme vulnerability of women engaged in coir work is exacerbated by the lack of alternate livelihood opportunities in the region, and their dependence on coir mill owners for raw material, i.e. coir fibre. An analysis of the value chain further reveals women's vulnerabilities in the coir industry.

Different stakeholders receive varying economic gains from the coir industry. The total profits made by each stakeholder are determined by two factors—per-unit profit margin and value transaction. Millers, fibre distributors, and collecting agents have a higher total income because of their high-volume transactions although they have a low percentage share in the whole value chain. Exporters enjoy an edge over all other stakeholders owing to good per-unit profit margins and high volume transactions. Men dominate these high-return roles and receive a larger share of the gains from the coir industry. Women who are mainly yarn spinners, yarn braiders, and home-based coir workers have low profit margins and small volume transactions. Thus, they are at the bottom of the value chain and are the most vulnerable in the coir industry.

6.5 Land Ownership

Most women engaged in coir work in the study area did not own even the land on which they lived. Women lived in small plots of lands owned by the men in the family. Most houses are clustered together, and the residential pattern reveals a close-knit kin network. The houses that were damaged were categorized as fully or partially damaged. Fully damaged houses were mainly in the Buffer Zone[3] but were rebuilt away from the coast. In all, 216 families in Peraliya North, 155 in

[2]A small and medium scale trader who provides women with coir fiber and buys back coir ropes woven by the women.

[3]The state imposed a ban on house construction within 100/200 m from the sea in the south and north east.

Peraliya South, and 34 in Thelwetta were relocated inland. As stated elsewhere (Goonesekera 2006; Kottegoda 2007; Perera 2010; Pinto-Jayawardena/de Almedia Gunarathne 2010), women in the study area may have lost their right to land when new houses were allocated to the heads of households rather than to title owners. Relocated families were not interviewed for this study as they are not currently engaged in coir work.

Those employed in the coir industry now live in the same houses that they lived in prior to the tsunami. The State provided compensation for repairing the damage caused to these houses. Compensation depended on the magnitude of damage and the money was dispersed in instalments according to the work completed. This compensation was given to heads of households rather than owners. Men are commonly accepted as heads of households unless they are physically absent from families. Some widows did own the land and house, but were marginalized due to this government action since they were not considered heads of households. In such cases, a male relative received the money. Some women created affidavits to empower male family members to collect the money on their behalf to avoid tedious bureaucratic processes. Women already had a weak entitlement to land, but following the tsunami, this was weakened further. State support after the tsunami came in many forms: cash compensation; land and house rehabilitation; and livelihood support. Women's rights were neglected in all three areas, with men receiving cash and land titles as head of household, and the fishing industry being restored while the coir industry was neglected. Thus, post-tsunami, women's income was also affected.

6.6 Coir Work in the Study Area and Impact of the Tsunami

Coir fibre extraction and spinning and processing of coir products have generated income for many people, especially women, on the southern coast of Sri Lanka for many years. In all, 222 persons were engaged in coir work in the study areas prior to the tsunami (Department of Census and Statistics 2005). Coir fibre was chiefly sourced from water pits and coconut fibre mills (*kohu mola*) from the surrounding GN divisions. These pits and coconut fibre mills (*kohu mola*) were owned by male entrepreneurs who also owned coconut groves as small holdings. The water pits were dug around water bodies in close vicinity of the study area or nearby lagoons. Some women worked in the coir pits emptying coconut husks left for 3–6 months to be fermented.

Coir fibre from the fermented husks was removed either manually or using a small machine. The manual process required women to hammer the soaked husks using a small wooden club until the fibre was loosened. The coconut fibre mills, semi-mechanized units, mainly employed men to put the wet husks through the crusher.[4] This task was considered dangerous and unsuitable for women. After the tsunami, coir fibre was also obtained from a women's NGO that operated in the south of the country. The main activity for women was spinning ropes from the coir fibre. Ropes were produced either by hand, by rubbing two strands between the palms, or by the more popular method of using a spinning set (machine). Three people, mostly women, were required to spin the machine. The finished products were used as ropes and woven as mats, carpets, and sacks.

Women lost their spinning sets (machines) when the tsunami waves swept into their houses. Coir pits, coconut fibre mills and other equipment was also washed away. Individual garden spaces where spinning took place were also damaged and trees uprooted. The statistics collected in February 2005 by the Department of Census and Statistics, revealed that though 59, 30 and 133 people were engaged in the coir industry in Peraliya North, Peraliya South and Thelwatta GN divisons of Galle district respectively before the tsunami, none of them engaged in the activity for about a year after the tsunami.

The destruction of the coir industry, which formed the chief livelihood of the area's women, was also highlighted in other studies (Oxfam 2006; Goonesekera 2009; Perera 2010). Women became more vulnerable as other income-generating activities such as sewing and making eatables for sale were also disrupted. Women also suffered as men lost their main livelihood of fishing.

6.7 The Recovery Process: Women Gain Ownership of Coir Fibres

After the tsunami, the immediate concern was to distribute relief in the form of food, clothing, and money and provide shelter. The government took primary responsibility for these tasks. However, a number of international, national, and local NGOs and other donors also helped provide relief and shelter to the affected people. In fact, the considerable assistance that poured into the area created dependency, jealousies, and ill feeling among people (UNDP 2005).

The most demanding task for the affected population was to restore lost livelihoods. The government's main concern was to re-establish the fishing and tourism industries, which were seen as important contributors to the national economy and the main livelihood for men who are considered breadwinners of families.[5]

[4]The crusher has a rotating wooden cylinder fitted with sharp steel spikes and the husk is held by hand by the operator.

[5]The State provided fishing boats and fishing gear to those registered with the government and having formal identification. Civil society organizations and individual donors, mostly foreign, also distributed fishing boats and fishing gear, often haphazardly, to people who did not even engage in fishing! International agencies, such as Food and Agriculture Organization, Canadian International Development, and ITDG South Asia pumped in money for the reconstruction of harbors, anchorages, and landing centers. FAO also assisted in preparing a master plan for fishing

In contrast, women's main source of livelihood did not receive the same attention and the government took no steps to assist in the restoration of this industry. Water pits were not re-dug due to environmental and health hazards. Coir work was carried out mostly as a home-based economic activity, and lacked official approval such as registration, license and insurance. Mostly low-income women were engaged in making ropes from coir fibre; they were not covered under insurance and thus could not claim damages. Some international and national NGOs helped by giving them coir fibre and spinning sets (machines).

A local women's organization came to the assistance of the women in the study area. The origins of this organization may be traced back to 1987, when small women's groups were started in villages under the influence of the "Change Agent Programme" introduced by the government. The 32 small groups operating in four villages joined together to form a women's organization in 1992; it was registered with the Department of Social Services. At present, the organization has 128 small societies across 17 villages with 1,590 women members. The organization is one of the 13 NGO members of the Galle district NGO consortium.

According to the organization's estimates, 1,006 families from 10 *community-based organizations* (CBOs) along the coastal belt lost everything due to the tsunami. Most of the members of these CBOs were engaged in coir-rope making and small businesses such as making eatables for sale, sewing, and petty trade. As one woman pointed out, "For 1 or 2 years, we could not start our coir work and we faced many economic hardships and we wanted very much to get back to weaving coir ropes." This need prompted the women to discuss their economic plight and look for solutions in their group meetings. These discussions transformed into a strong demand for a coir-processing centre. A woman in a focus group discussion noted, "We all got together and asked for a *kohu mola* (coir mill) because we were ready to start weaving coir ropes if we had coir fibre."

Subsequently, with participation from local women, a coir-processing centre was set up in Peraliya by the women's organization in 2007. The plant had adequate capacity, stores, a drying yard, and office space to serve 175 tsunami affected families and 150 families outside the three villages. The centre was financed mostly by a national NGO that funds participatory development activities through district NGO consortiums. It also made cash grants and gifted spinning sets (machines) for the women. The women's organization too made a financial contribution and formed a management committee to maintain and develop the coir-processing centre and empower the women engaged in coir work. These women were also trained in marketing and management techniques. With the opening of the centre, the women overcame the difficulty of finding fibre and were able to restore their

(Footnote 5 continued)
infrastructure reconstruction and development. By 2007, the industry had recovered from the loss, according to the Government *Reconstruction and Development Agency* (RADA): of the "over 19,000 boats destroyed, 90 % had been replaced and fishermen were now catching 70 % of what they did prior to tsunami" (IRIN 2007). Fishermen in the study area now use multi-day boats, which many did not have earlier, and they spend a number of days at sea.

livelihood, demonstrating their resilience and capacity to overcome disasters through collective action. More importantly, the women gained a certain degree of pride as owners of the raw material since they no longer had to depend on *mudalalis* (small businessmen) as was the case before the tsunami. As one woman said, "Now we are not under the *mudalalis* as we have our own *kohu mola* (coir mill)." Whenever the women mentioned the coir-processing centre, they referred to it as "our own *kohu mola*".

The centre functions as a company where each member of the local women's organization buys as many shares as she can afford to. At the end of the year, dividends are distributed according to profits gained over the year. A man (relative of a member) was hired on a monthly wage to collect coconut husks from different places, transport them to the centre, and run the coir machine. The coir centre had good demand from women for coir fibre. Women even had to queue to get their coir fibre, but they did not mind it as they received coir fibre at a lower price. Delivery of coir fibre to individual homes was undertaken for an additional charge. The women also had the advantage of selling coconut husks collected from their gardens to the processing centre. The extra money earned was used to purchase coir fibre from the processing centre.

Women spin ropes in small teams of three. While one woman turns the wheel, the other two keep walking backwards in a circle, spinning. One woman stated, "In 1 day, we walk a distance similar to walking from Colombo to Galle (123 km)." Although the task is tedious, women talk to each other while working, share their experiences and difficulties, and help each other in times of need. There is a close bond among the women as most are either related or neighbours. The collective work creates solidarity among the members. Generally, women start weaving at about 8.30 a.m. and stop for lunch at around 11.00 a.m. This gives them time to attend to domestic chores. After lunch, they weave from 2.00 p.m. to 5.00–6.00 p. m. They usually weave three bundles of coir skeins per day worth around Rs. 300.

The organization further provided vocational training to economically empower women. Women were trained to make brooms, brushes, mats, carpets, etc., using coir fibre and other materials obtained from coconut trees; this helped women secure extra income. However, some of the women who were trained in this fashion have given up such work due to the difficulty in obtaining raw material. The income from coir and related work remains an important source of family income. One woman observed, "When our men (husbands) do not go to sea, this is the only income (coir work) we have". During certain months of the year, the seas are too rough for the men to venture out for fishing.

Members of the women's organization also help the women save and access credit. Every member is required to save a few rupees (Rs. 5–10) a week; this is deposited in a bank account maintained by the organization. The members are also entitled to credit at a very low interest rate. Members have not only benefited economically but have also been able to develop social capital.

Initially, the members of smaller groups met once a week, but after the tsunami they could meet only once a month. If the need arises though, they meet anytime before the monthly meeting. At these meetings, women discuss matters pertaining to their livelihoods and their families. They help each other as and when the need arises. They also have informal savings (*seettu*) which are popular among low-income groups, especially women.

In the study villages, women did not feel domestic violence was an issue. In many families, the husband and wife jointly make decisions. However, finances are managed by women; this increases their burden of fulfilling family. For some women, this is a source of pride. One woman noted, "I am the strength, the force in my family. Neighbours know the contribution that I make to build my family."

6.8 Rebuilding the Coir Value Chain

A few years after starting the coir-processing centre, they faced a major problem since some of their employees migrated and replacements were difficult to find. Collecting husks from different places and transporting them to the processing centre was also a problem for women, although they were efficient in other tasks. The patriarchal practices of limiting women's mobility and their access to different skills meant that women had little experience collecting coconut husks from different places and transporting them to the processing centre. The centre had to be closed for a few months as the organization could not find a male worker willing to collect husks. Men preferred to undertake other economic activities that offer better remuneration.

The women then came up with an alternative strategy: a committee member and her husband took on the responsibility of collecting husks and producing coconut fibre at the centre. Instead of paying the couple a wage, they were given a proportionate share in the profit. Innovative as this strategy is, other problems have arisen. The couple runs a restaurant that attracts many customers. Therefore, they do not have enough time for the coir mill, and providing coir for the requirements of the members has become a challenge for the organization.

This led to another innovation. Another women's NGO, operational prior to the tsunami, has come to the aid of the women, and now provides coir fibre for the women coir workers. This arrangement was not difficult as both organizations are members of the district NGO consortium where NGOs work collaboratively. Poor women gain some advantage since this NGO gives coir fibre on credit and women give them woven ropes in return. The organization pays Rs. 250 for one bundle of coir skeins of which Rs. 50 is retained as savings that women can access at the end of the year. These compulsory savings are highly valued by the women who use them for different family and personal needs.

Globalization trends and the market economy are powerful forces that have become a threat to the sustainability of women's main livelihood. Today, coir ropes have to compete with synthetic materials in the market. Women's groups are also competing with *mudalalis* (small businessmen) for raw material and the sale of finished products. Women coir workers also have to compete with large companies that export high-quality fibre and coir products for raw material. Women complain that they do not get a fair price for their products in the market, which impacts them negatively. Although the State is responsible for promoting and revitalizing the home-based coir industry, the indifferent attitude of the State is harmful to women who form the majority of its labor force. As Sarala Emmanuel says, "The responsibility [to ensure that women workers get a fair deal] does not only lie with women's groups or networks, but with the Sri Lankan state and all organizations that have objectives of social justice, equal rights, and women's empowerment" (2009: 166).

6.9 Conclusion

The tsunami destroyed livelihoods and the State and NGO sector focused on fishing which employed 40 % of the men, ignoring coir which employed 35 % of the women in the study area. In fact, the quantity of aid given to fisheries sector was so lucrative, that local businessmen benefited enormously from it and did not consider the coir industry worth reviving. Thus, coir formed a kind of invisible space in the study area after the tsunami.

Women had weaker claim to land in the study area, while external aid has further strengthened men's land ownership and supported the reconstruction of their livelihoods through land ownership. Hence coir industry did not get much attention, and women had more difficulty while rebuilding their livelihood after tsunami. This chapter demonstrated how despite such adversities, women have collectively revived the coir industry and through their network have actually transformed the coir value chain for their benefit.

Factors that lead to the successful recovery of the coir industry include women's collective strength, local women's organization, CBOs, NGO consortium, and a national level NGO. Together, they set up a coir-processing centre, and women took on the responsibility of implementing and managing the centre. This gave them the opportunity to own raw materials for the first time, for which they had earlier depended on *mudalalis* (small businessmen). Networking with other NGO members of the district NGO consortium allows women to receive coir fibre from another local women's organization when they run short, further developing women's collective strength. In addition to reviving their livelihood, women have been able to save and receive credit through their collective activities. They have created space for themselves at regular meetings to discuss their day-to-day problems with one another, and find solutions. The trainings they received have made them more confident and helped them realize their strength as a group to deal with oppression in society, especially domestic violence. They feel that they are not alone and can turn to other women for help.

Women's own agency led to the revival of the coir industry and created new value chains that were uniquely suited to women's needs. Despite the gender blindness of the officials of both the State and the relief and rehabilitation industry, women rebuilt their lives and livelihoods. Not just this, they also revived a dead industry, which was ignored by all actors in the post-tsunami reconstruction. To rebuild their own livelihoods, women had to revive areas higher up the value chain through their networks with other women's groups. This is an example of women's agency stretching all the way from their own homes to encompass the entire industry in the study area.

Disasters are essentially social phenomena with foundations in social structures that are highly stratified with numerous gender, class, and ethnic divisions (Kapadia 2013). Disasters further weaken women's land rights. Since external assistance is done through head of households and land owners, women find it difficult to rebuild their livelihoods. The present study clearly illustrates the extreme vulnerability of women from low-income groups to disasters. However, the tsunami allowed women to overcome the exploitative relationships that characterized coir-fibre procurement before the disaster. By establishing their own coir-processing centre, women's collective strength was a powerful force that erased their dependency on male entrepreneurs who controlled the sale of coir fibre and purchase of coir products. However, powerful patriarchal practices limiting women's mobility and acquisition of skills continue to limit women's agency, making them dependent on a male worker to collect and transport coconut husks to the centre; this threatens the centre's sustainability. In addition, globalization trends and market forces threaten the home-based coir work of women from low-income groups in coastal areas. Thus, continuous support from the State and donor agencies is needed for women to face such challenges.

References

de Mel, Neloufer, 2007: "Between the War and the Sea", in: *Interventions*, 9,2: 238–254.

de Mel, Neloufer; Ruwanpura, Kanchana N., 2006: *Gendering the Tsunami: Women's Experiences from Sri Lanka* (Colombo: ICES).

de Mel, Neloufer; Ruwanpura, Kanchana N.; Samarasinghe, Gameela (Eds.), 2009: *After the Waves: The Impact of the Tsunami on Women in Sri Lanka* (Colombo: Social Scientist's Association).

Department of Census and Statistics, 2005: *Preliminary Statistics of the Census of Buildings and Persons Affected by the Tsunami—2004.* (Galle District. Colombo: Department of Census and Statistics).

Emmanuel, Sarala, 2009: "The Post-disaster Development Transition: An Analysis of Community–Level Women's Mobilization, Citizenship and 'Good' Governance", in: de Mel, Neloufer; Ruwanpura, Kanchana N.; Samarasinghe, Gameela (Eds.): *After the Waves: The Impact of the Tsunami on Women in Sri Lanka* (Colombo: Social Scientist's Association): 153–178.

Fothergill, Alice, 1996: "The Neglect of Gender in Disaster Work: An Overview of the Literature", in: *International Journal of Mass Emergencies and Disasters*, 14,1.

Goonesekera, Savitri, 2006: *A Gender Analysis of Tsunami Impact: Relief Recovery and Reconstruction in Some Districts in Sri Lanka* (Colombo: Centre for Women's Research).

Goonesekera, Savitri, 2009: "A Gender Analysis of Tsunami Impact: Relief Recovery and Reconstruction in Some Districts in Sri Lanka", in: de Mel, Neloufer; Ruwanpura, Kanchana N.; Samarasinghe, Gameela (Eds.): *After the Waves: The Impact of the Tsunami on Women in Sri Lanka* (Colombo: Social Scientist's Association): 84–111.

Hyndman, Jennifer, 2009: "Troubling 'Widows' in Post-tsunami Sri Lanka: Feminist Meets Double Disasters," in: de Mel, Neloufer; Ruwanpura, Kanchana N.; Samarasinghe, Gameela (Eds.): *After the Waves: The Impact of the Tsunami on Women in Sri Lanka* (Colombo: Social Scientist's Association): 17–41.

IRIN, 2007: "Sri Lanka: Fishing Industry Still Lacks Large Tonnage Boats", 21 June; at: http://www.irinnews.org/printreport.aspx?reportid=72857: p. 1.

Kapadia, Kamal, 2013: "Producing Entrepreneurs in Sri Lanka's Post-Tsunami Economy: Re-thinking the Relationships between Aid, knowledge and Power", in: *Contemporary South Asia*, 21,1: 6–21.

Kottegoda, Sepali, 2007: "In the Aftermath of the Tsunami Disaster: Gender Identities in Sri Lanka", in: *Women in Action*, 2: 16–24.

Kumara, A.G.L.A.S.; Hirimuthugoda, N.Y.; Edirisinghe, U., 2011: "Assessment of Socio-economic and Health Status of Fisherwomen Involved in Coir Retting in Dickwella Secretariat Division, Matara District", in: *Tropical Agricultural Research*, 23,1: 84–90.

Mubarak-Perera, Kamakshi, 2013: "Positive Responses, Uneven Experiences: Intersections of Gender, Ethnicity and Location in Post-tsunami Sri Lanka", in: *Gender, Place and Culture*, 20,5: 664–685.

Oxfam, 2006: *The Coir Industry in the Southern Province of Sri Lanka* (Colombo, Oxfam).

Peiris, Kamala, 1997: *Weaving a Future Together* (Utrecht, The Netherlands: Institute for Development Research).

Perera, Sriyani, 2010: *Women's Land Rights in Post-disaster Contexts* (Colombo: We Women Against Violence Network).

Pinto-Jayawardena, Kishali; de Almedia Gunarathne, Jayantha, 2010: *Is Land Just for Men? Critiquing Discriminatory Laws, Regulations and Administrative Practices Relating to Land and Property Rights of Women in Sri Lanka* (Colombo: Law & Society Trust).

Risseeuw, Carla, 1980: *The Wrong End of the Rope: Women Coir Workers in Sri Lanka* (Leiden: Research Project Women and Development).

Ruwanpura, Kanchana N., 2007: "Awareness and Action: The Ethno-gender Dynamics of Sri Lankan NGOs", in: *Gender, Place and Culture*, 14,3: 317–333.

Ruwanpura, Kanchana N., 2008: "Temporality of Disasters: The Politics of Women's Livelihoods 'After' the 2004 Tsunami in Sri Lanka", in: *Singapore Journal of Tropical Geography*, 29,3: 325–340.

Ruwanpura, Kanchana N., 2009: "Waves of Reality: Spatial and Temporal Dynamics of Post-tsunami Development Efforts in Sri Lanka—A Gender Analysis", in: de Mel, Neloufer; Ruwanpura, Kanchana N.; Samarasinghe, Gameela (Eds.): *After the Waves: The Impact of the Tsunami on Women in Sri Lanka* (Colombo: Social Scientist's Association): 65–83.

Silva, Kalinga T., 2010: "Are Women Merely Passive Victims of Disaster or Active Agents in the Recovery from Disasters? The Case of Tsunami in Sri Lanka", in: CENWOR (Ed.): *Charting Pathways to Gender Equality: Reflections and Challenges* (Colombo: Centre for Women's Research): 159–170.

UNDP, 2005: *The Report on People's Consultations on Post Tsunami Relief, Reconstruction and Rehabilitation in Sri Lanka,* (Colombo: UNDP, Disaster Relief and Monitoring Unit—Human Rights Commission Sri Lanka and Colombo: University Extension Centre).

Chapter 7
Gender, Land Tenure, and Disasters in the Mentawai Islands, Indonesia

Carol Yong, Frans R. Siahaan and Andreas Burghofer

Abstract This case study discusses the many challenges faced by the Mentawai people of Indonesia who are increasingly suffering the consequences of unsustainable development and complex natural and human-made disasters. A brief review of the rich biodiversity and natural resources of Mentawai Islands contextualizes the discussion that follows, namely the different threats to lands, forests and natural resources the Mentawai people find themselves grappling with. This chapter identifies a link between gender, natural resource, land tenure, and disaster and argues that nuances around gender and other categories are inadequately addressed or ignored. We focus particularly on three issues: first, locating Mentawai women's inheritance rights within the Mentawaian patrilineal social structures and customs that circumscribe women's rights to land and other property; second, the connection between women's land tenure and gender differential disaster damages and recovery process; and the third, how gendered land rights/tenure interrelate with the difference of loss and suffering of women and men in Mentawai society and how development agendas and interventions construct vulnerability of women and men differently.

Keywords Gender · Land tenure · Disaster vulnerability · Indigenous people · Mentawai

Abbreviations

HPH	Hak Pengusahaan Hutan
IPK	Izin Pemanfaatan Kayu
WALHI	Wahana Lingkungan Hidup Indonesia
YCM	Yayasan Citra Mandiri

Carol Yong, independent researcher; Email: yrn@vjfg.at; Frans R. Siahaan, Program Officer, The Asia Foundation; Email: frans.siahaan@gmail.com; Andreas Burghofer, Director Upper Austrian branch office of Gesellschaft fuer bedrohte Voelker (GfbV); Email: andreas.burghofer@gfbv.at.

K. Kusakabe et al. (eds.), *Gender and Land Tenure in the Context of Disaster in Asia*, SpringerBriefs in Environment, Security, Development and Peace, DOI 10.1007/978-3-319-16616-2_7

7.1 Introduction[1]

The Mentawai archipelago is located about 120 kms off the western coast of Sumatra, Indonesia. The archipelago is made up of over 40 small islands and numerous islets in about 6,011.35 km^2. Only four islands are relatively large and inhabited—Siberut (3,850.16 km^2), South Pagai (879.61 km^2), North Pagai (602.43 km^2), and Sipora (594.70 km^2)—and are together known as Mentawai Islands (Bappeda/BPS 2011). The islands have a tropical monsoon climate with 4,500–6,000 mm of rainfall per year.

Mentawaians have faced ethnic discrimination and economic exploitation due to the adverse political climate in the Minangkabau-dominated province of West Sumatra. The main problem has been the lack of recognition and protection of Mentawai indigenous identity and their economic, social, political, and cultural rights by the government and dominant sections of society. Additionally, government-endorsed large-scale exploitation of forests for timber has resulted in deforestation and further loss of rights to customary lands. The Mentawaians live in a region that is highly prone to natural disasters, because of its geographical location along Sumatra's major earthquake fault line.[2] The soil on Mentawai Islands is very fragile and has no solid rock base. It is easily degraded and washed away. Logging, oil palm plantations, introduction of pests and diseases, and extensive land use changes significantly increase the vulnerability of the islands and their inhabitants.

Our approach is to develop grounded understanding and insights into disaster risks to help enrich gender analysis and offer practical solutions according to local Mentawai perspectives, experiences, and priorities. This case study is based on a variety of enquiry methods and approaches. This includes literature review; material gathered through interviews, focus groups and observations during field visits to Mentawai Islands, and on the authors' own practical working experience and knowledge of the issues in Mentawai Islands/Indonesia and others, for example, gender, human rights, indigenous people and land rights. It is important to mention that the second author has been working for 11 years with a local non-governmental organization and living in Mentawai rural villages over the years. This long-term engagement provided an insider's lens into the Mentawai situation, issues and changes facing Mentawai people at different times and administrative status, from local perspectives. Discussions and analysis was also derived from earlier field visits and stay in Mentawai rural villages, of two authors in the Mentawai

[1]Special thanks are due to Yayasan Citra Mandiri (YCM), Puailiggoubat and the Mentawai people for sharing their stories and experiences with us during the field visits, and for allowing us to write about them. We are grateful to the anonymous reviewers who commented on draft versions of this case study.

[2]The Indonesia archipelago is itself located on the 'ring of fire' making it one of the most disaster-prone regions in the world.

Islands—one as consultant and the other as NGO resource.[3] This chapter also draws on the rich documentations of *Yayasan Citra Mandiri* (YCM) and other non-governmental and governmental sources and on-going discussions between ourselves and other NGO colleagues in Indonesia.

7.2 Background and Context: The People of Mentawai Islands

The Development Planning Agency and Central Board of Statistics (Bapedda/BPS 2011) of Mentawai Islands Regency estimated the population of Mentawai Islands at 76,173 people in 2011 with 36,699 women and 39,504 men and a density of about 12 people per km^2. The main inhabitants are indigenous Mentawaians. They belong to the Austronesian language family. They tend to live in remote areas of the islands deep in the rainforest, near a river and sago grove. Traditionally, they use dugout canoes, called *pong-pong* or *abak* to travel along the rivers. When outboard motors came, those with money acquired motor-operated boats.

Mentawaians are organized into groups of patrilineal kin or clans and live in traditional communal houses called *uma*, found alone or in small clusters. *Uma* is also the highest level of Mentawaian social and political institutions. Clans of one or more *uma* form the village, locally called *laggai*, although the current government still uses *desa,* an Indonesian word which means a village of several households that doesn't reflect the clan system of the Mentawaians (YCM 2007).

Geographical isolation and harsh physical conditions helped preserve the Mentawai identity and way of life. Over time, however, the Mentawaians lost much of their identity and traditional practices. For example, the Mentawai traditional religion or belief system known as *Arat Sabulungan* was banned by the Indonesian government in 1955.[4] Another example is the old tradition of body tattoos that are now only seen on older women and men (authors' field observations and personal conversations with tattooed elders, 2007, 2013).

Mentawaians still follow the traditional practice of hunting and gathering in the forests that provides them with medicines and material and non-material resources used in rituals performed by the *Sikerei* (traditional healer/shaman). Clusters of sago palms in the marshy lowlands form the staple diet of the Mentawaians. Cassava,

---

[3]The aim of the two field visits was principally for the external evaluation of the YCM, using an integrated approach and methods to collect both qualitative and quantitative data in Mentawai villages over several weeks. It should be clear that the evaluation was not for this publication. With permission from YCM and Puailliggobat, we adapted some of the information collected during the field visits for this chapter, and we give special thanks to them and their funder.

[4]The Mentawaians, like other Indonesians, were only allowed to adopt one of five official religions: Protestantism, Catholicism, Islam, Hinduism, or Buddhism. Most Mentawaians are Protestant; Catholicism ranks second, followed by Islam.

taro, coconut, banana, and other fruit trees are cultivated around the *uma* (authors' field observations; Hernawati 2004). Most families also keep a few pigs and chickens, but fish is an important source of protein in their diet. Women and children catch fish, shellfish, shrimp, etc., from the rivers and marsh forests. Some fish are dried and stored for the lean or drought seasons. Rivers are important for transportation and demarcating boundaries between the lands and forest areas of each *uma*. The forest and farmland is usually quite far, so the family normally stays several days or weeks in the farm hut. Traditional economic activities are performed collectively.

Despite staying in remote parts of the islands, Mentawaians had contact with outsiders including British and later Dutch colonial officers, explorers, traders, missionaries, and occasional researchers. Mentawaians collected and sold rattan, patchouli, and cloves to the Minangkabau visiting the sub district centres of Mentawai Islands for trade. The Minangkabau monopolized the economy and politics of West Sumatra and the Mentawai Islands. Some eventually settled on the islands, because transportation and communications between Padang and the islands were often disrupted by bad weather. During the rainy season, between June and September, strong rainfall (accompanied by thunder) and overpowering waves make sea travel almost impossible. Today, the Minangkabau form the bulk of the non-indigenous population in the Mentawai Islands. Other migrants consisted of the Batak and Nias from North Sumatra, Javanese and other Indonesian ethnic groups; others were transferred to the islands as civil servants. The migrant population mainly established itself in the coastal villages of the district and sub district centres, as traders, contractors, shopkeepers, or café owners.

The Mentawai Islands became more accessible after gaining autonomy in 1999 under Law No. 49/1999 (Undang-Undang Republik Indonesia Nomor 49/1999) on the Establishment of Mentawai District. The district government began developing infrastructure and services and tourist facilities in Tuapeijat, on Sipora Island, the capital of Mentawai District (Office of Tourism, Arts & Culture of Mentawai, 2004). This brought in more outsiders who wanted to stay close to the ports to access transportation and other services. Local people were also attracted to the centres by promises of jobs, health services, education, etc. The image of the Mentawai Islands as a surfing paradise with adventure tours of exotic, dense tropical rainforests was another draw. Outsiders introduced commodities such as bread, rice, canned food, sugar, and biscuits to the Mentawai Islands; today, these have become part of the household diet in many Mentawai villages. The pressure and impact of these changes on the fragile Mentawai forest ecosystem and environment is enormous, and is further exacerbated by haphazard and poor town planning. However, Mentawai women and girls are still largely absent from decision-making that affects their lives and have limited participation in customary land, education, health, and public policy.

7.3 Biodiversity and Natural Resources

The Mentawai Islands were once rich in natural resources. Almost 75 per cent of these islands was covered in tropical rainforests with high biodiversity, particularly Siberut, the largest island in the Mentawai archipelago. However, much of Siberut's forests have been destroyed since the 1970s. Two main factors drive deforestation in Mentawai: (1) legal logging via large-scale forest concessions (*Hak Pengusahaan Hutan*, HPH) and small-scale forest concessions or timber utilization permits (*Izin Pemanfaatan Kayu*, IPK) and (2) illegal logging (clearing forests without permits). Both activities are conducted by logging and plantation companies.

Economic expansion through natural resource exploitation was justified since it was expected to bring prosperity to Mentawai district. Yet, investigations by YCM revealed that the forestry sector's contribution to the locally-generated revenues (*Pendapatan Asli Daerah*/PAD) for 2004 was zero (YCM 2006). Additionally, YCM and other non-governmental organizations (NGOs) exposed cases of massive corruption; misuse of development funds; legal and illegal logging in Sikakap, Sipora, and Siberut; planned conversion of some 70,000 hectares of land in Siberut for commercial oil palm plantations; and land disputes between local communities and local government officials/companies (YCM et al. 2004).[5] In effect, the natural resource management policy lacked strict regulations on logging and on forest management by local governments.

Uncontrolled commercial logging, legal and illegal, adversely threatened the biodiversity and natural resources of the Mentawai Islands. Since early 1969, five HPHs have been granted to exploit Siberut Island's forest, which covered over 285,000 hectares. Logging was widespread and extensive, especially in Siberut Island, from 1972 to 1993, dubbed the "first phase" of large-scale logging. Darmanto/Setyowati (2012) estimated that 30,650 hectares of Siberut's forests were logged for timber production during those years while Munawar (2004, cited in Darmanto/Setyowati 2012: 155) calculated that some 1.7 million cubic meters were logged. The former Indonesian President, Soeharto, issued a decree in March 1992 to stop all logging in Siberut Island (Barber et al. 1997: 75). This was in response to the central government initiative to establish the Siberut National Park funded by the *Asian Development Bank* (ADB).

With decentralization, the Mentawai regency government assumed responsibility for managing the islands' forests and other natural resources. Like other governments of forest-rich districts across Indonesia, the Mentawai District Government responded rapidly by issuing a District Head's Decree and a District Regulation to control locally issued IPKs. Permits to log natural forests under the IPK scheme had accelerated the exploitation of forest resources. The Mentawai District Government

[5]One high profile corruption case in the sector involved the former regent and the former head of the forestry department, who were sentenced to 4 and 5 years in prison, respectively.

Table 7.1 Forest concessions issued in Siberut island

No.	Logging companies	Areas (ha)	Concession types	Begins	Ends
1	PT. Cahaya Pharmin Siberut	35,000	HPH	1969	unknown
2	PT. Cirebon Agung	70,000	HPH	1970	1992
3	PT. Kayu Siberut Inkappa	110,000	HPH	1973	1992
4	PT. Sumber Jaya Indah	70,000	HPH	1977	1992
5	PT. Sumber Surya Semesta	unknown	HPH	1987	1992
6	Koperasi Andalas Madani	45,650	HPH	2001	2046
7	KUD Sikabaluan	1,000	IPK	2003	2006
8	KUD Sikabaluan (IPK Jalan)	1,664	IPK	2003	2007
9	KSU Taileleu Mandiri	1,000	IPK	2003	2006
10	KSU Purimanuaijat	700	IPK	2003	2005
11	PT. Alam Indah Lestari	1,000	IPK	2004	2007
12	KSU Mitra Sakato	1,800	IPK	2004	2007
13	PT. Suma Salaki Sejahtera	49,440	HPH	2008	2052

Note HPH is legal logging via large-scale forest concessions (*Hak Pengusahaan Hutan*) and IPK is small-scale forest concessions or timber utilization permits (*Izin Pemanfaatan Kayu*)
Source Siahaan (2009)

regarded forest resources (timber) as one of the most valuable and important resources that could help increase locally-generated revenues (PAD). Until 2008, it had granted 30 IPKs across the district for 31,244 hectares and had recommended four HPHs to the central government to harvest Mentawai's forest. From 2001 to 2013, the District Head granted seven IPKs in Siberut Island covering 6,164 hectares, while the national-level Minister of Forestry in Jakarta granted two HPHs covering 95,090 hectares (See Table 7.1).

According to UNESCO (2004, 2005), "Active and pending logging concessions threaten to destroy, at a minimum, 70 % of the natural habitat remaining outside Siberut National Park, and are likely to impact an even greater area including the interior of the National Park." The scale and severity of uncontrolled logging, both legal and illegal, is a major problem. Massive deforestation has direct and indirect effects on the biodiversity, natural resources, and forest ecosystem of the Mentawai Islands. Encroachment and forest degradation affects the Mentawaians' rights to lands, forests, and resources, including access to resources for community elders who are traditional healers and ritual caretakers. Diminishing water resources, especially in logging areas, exert fresh pressure on women and girls who often bear the main responsibility for food security for the family and for procuring water for washing, drinking, and cooking. Depleted forests and forest resources deprive Mentawai children of their playground and 'school'. Mentawaians reiterate that the government simply ignored the impact of logging operations on the forests and natural resources that they depend on for their livelihood and cash income and their socio-cultural, political, and spiritual way of life.

The oil palm industry cannot operate in Mentawai now because permits have expired and have not been renewed; two licenses were recently cancelled.[6] However, the threats to the forests remain. Oil palm companies are still lobbying the District Head for licenses and coaxing Mentawaians to accept plantations.

Logging, oil palm plantations, and commercial crops have adverse effects on forests' flora, fauna, and resource biodiversity. Some endemic species are already on the brink of extinction. The degradation and destruction of Mentawai's mangroves, peat swamps, beach forests, coastal ecosystems, and coral reefs—all unique forests and natural resources—have been increasing rapidly. These resources acted as buffer zones protecting the islands from the effects of climate change and natural disaster such as tsunami and abrasion. Their destruction increases the vulnerability of the islands and its people to natural disasters. These activities are transforming the landscape of the Mentawai Islands from dense tropical rainforests to barren land or mono-culture plantations void of biodiversity.

The activities that are destroying Mentawai Islands' natural resources are also attracting outsiders to the islands for work or business. Outsiders from mainland Sumatra and other overpopulated regions such as Java have also added to the pressure on the islands. The WWF and IUCN (undated) noted that: "The primary forests of the Mentawai Islands remained essentially intact until the influx of settlers from mainland Sumatra created population pressure and disrupted traditional management practices. Extensive areas of land have been cleared for cash crops, and logging has become a serious threat."

The increasing population in the islands creates a need for housing close to the centres and easy access to road transportation and port facilities. As a result, land traffic in areas close to the ports and shores also increases. Yet, with the Mentawai government's limited resources and expertise, effective infrastructure and settlement development are lacking. More tourists meant the construction of more roads, hotels, resorts, and other facilities, exerting further pressure on the land, resources, and local people. Both the construction and tourism businesses are largely dominated by outsiders.

In Mentawai district and sub districts, extensive stretches of land have been cleared for building roads, settlements, and other infrastructure. While development is necessary for a young autonomous region, changes have been rather rapid and without proper rules. Construction activities, for example, do not rely on local resources but use cement and other modern materials that can be harmful to the environment and local economies. An even bigger problem is construction by outsiders using coral gravel/sands. This degrades or destroys coral reefs, intensifying the islands' vulnerability to natural disasters, as was evident when the tsunami struck the islands in 2004. Experts have inferred from records of sea-level changes in the corals of the Mentawai Islands that the Sunda megathrust could trigger great earthquakes. They point out that the magnitude-8.4 earthquake of September 2007

---

[6]This information is drawn from the external evaluation of YCM unrelated to this case study, and we thank YCM for allowing us to include this.

"represents the first in a series of large partial failures of the Mentawai section that will probably be completed within the next several decades" (Sieh et al., 2008).

Land use problems are not restricted to district and sub district capitals. In many Mentawai villages, many food crops have been replaced or inter-cropped with cash crops, particularly cocoa. Space around sago palms by the Silaoionan riverside has been taken up by cocoa; in some cases, sago palms have been replaced by cocoa. Unsustainable farming systems degrade soil fertility and weaken traditional systems *uma*- and *laggai*-based land management.

7.4 Women's Rights to Land, Property and Inheritance in Mentawai Society

Although the Mentawai indigenous *uma*- and *laggai*-based land management has the value of preserving the security of land and sustainable use, it is also clearly gendered, especially the allocation of land rights and resources to women. In traditional Mentawai society, based on the patrilineal kin or clan system, women and girls are disadvantaged in access to land and property inheritance, compared with men and boys (Hernawati 2007: 60). For example, Mentawaian land tenure arrangements and cultural practices allow women and girls to use land and forests for collecting forest resources or cultivation but exclude them from access to and control over land in their own right. In addition to the lack of rights to inherit land, women are also disadvantaged in access, control and authority over most household property and children. Women can only inherit some ancestral items like cooking pots that they receive as wedding gifts.

During a focus group discussion in April 2013, some women felt that the traditional tenure of Mentawai culture must be changed, but others were pessimistic about women gaining land. The scope for change is limited, because single mothers, widowed or divorced women could even lose their children to male relatives. One woman pointed out that it is difficult for women to challenge the dominant view of men on tradition and custom that "legitimized" Mentawaian patrilineal descent and dictate property rights and care of offspring to be transferred along the male lineage, exclusively to men.

It is crucial to note that neither women nor men are homogenous, due to factors such as age, marital status, kinship role and status, and so on, resulting in different priorities and notions of land ownership. Since the land tenure system is grounded in the male-line kin, women's access is mainly through men, husbands or fathers. This can be expected to have important implications for women and other dependent household members, especially in the aftermath of disaster (discussed below).

Traditional land rights are ascribed to households, specifically men. But under certain conditions, women in fact can hold limited land and decision-making power on how to use such land. Symbolically, individual women can demand changes or find ways to persuade men to change their attitude and allow women to hold land.

For example, women can be given a plot of family land to grow taro (*keladi*). Women and girls can make decisions concerning this plot and these are recognized by the family and community, mainly because taro is culturally and economically considered a 'female' product, and consequently some women could gain access to lands by asking for specific plots to grow taro.[7]

The recent push to growing cocoa associated with 'modernizing' agricultural land use, as noted above, means that women are likely to be disadvantaged when there is less land available to be shared within the household. Women claimed that the decision to grow cocoa was taken mainly by men, although some may have consulted their wives. Women in a village of Central Siberut stressed that men usually controlled the income generated from cocoa. Some of them said they negotiated with their husbands for a share of the income to buy food for the household "because if they don't, that means no food on the table." This demonstrates that women seek ways to have some decision-making power in the household through their gender roles of cooking, meeting family needs, and so on.[8]

7.5 Gendered Processes of Disaster Damages, Response and Recovery

The lack of women's rights to land is a major issue for marginalized groups like the Mentawai people living in highly vulnerable and disaster-prone contexts. Land is an important asset, particularly as the means to secure livelihoods and ensure survival. Women need secure access to land rights/tenure if the impacts of natural, and increasingly man-made disasters are to be mitigated in a manner that will reduce women's dependence on men and humanitarian assistance. In addition, disaster-affected communities will not be able to adequately recover from disaster losses unless their access to land, and more specifically land rights for poor women and men, is taken into account. As Wisner et al. (1994, cited in Fordham 1998:127) said, "Disaster vulnerability cannot be separated from vulnerability in everyday living and this is seen most clearly through unequal access to resources."

Another important aspect of the ability to recover from disaster damages and rebuild lives is the level of support and access to key natural resources such as land, water, and trees. The barriers that restrict women from securing lands have underlying consequences on access to information and humanitarian assistance. If women do not hold land titles, then officials providing information on disaster

[7]However, we need to point out that this information was obtained exclusively from discussions with local Mentawai women in Siberut, and supported by the YCM field staffs. We suggest that further research is required, and more comprehensively, in order to identity such special arrangements for Mentawai women to access land and land use through cultivation of specific crops.

[8]More research is needed to explain gender processes operating at local levels in the household and in communities affected by market-based options of land use.

mitigation or precautions usually do not address them. This is disadvantageous for women-headed households as well as family members who depend on women, such as children. For example, when a school was flooded, women did not know or learned too late that relief was paid to the school as well as affected children for the losses suffered.

Disaster-affected communities cope with and recover from disaster in multiple ways, as in the case of flood-stricken Siberut Island (see box). But disaster response and recovery is gendered and intersect with factors such as age, status, kinship role, etc. (Cupples 2007; Enarson 1998; Fordham 1998; Fulu 2007). For instance, aid and other associated benefits for victims were passed to village headmen, household heads or community representatives. Thus mostly men received them. Similarly, younger women and men also collected emergency food packages and other hand-outs, because they owned motorbikes or were physically fit to walk to the relief centres. Those with limited resources, including lack of transport and living far from the *posko* (relief base camp) can experience particular difficulties to avail disaster relief assistance and services.

Organized communities in areas serviced by NGOs such as YCM appeared more effective and resilient in post-disaster reconstruction. They were able to organize themselves to contribute labour to carry out construction work of repairing broken bridges and roads, and building new houses. The local government provided the villagers and households some plywood, zinc, and so on for the repairs. These initiatives were however led or dominated by a group of men and youth. Some women and older men do the simpler repairs and tasks of cleaning up floating debris, salvaging scattered household objects or school facilities such as chairs, tables, books, etc.

When the government decided to determine temporary housing locations for tsunami refugees without consulting landowners, the situation worsened, causing conflicts between landowners and refugees (Puailiggoubat 2013), reinforcing the refugees' vulnerability. Conflicts can make women and girls more vulnerable in male-dominated spaces. Some NGO activists providing emergency services reported whispers of rape and sexual violence against some women and girls in shelters. Such cases were unconfirmed because the victims were afraid to testify, owing to their dependency on humanitarian aid from outside authorities and agencies.

Box 7.1 Case study: Floods of April 2013

In April 2013, continuous rains triggered unprecedented flooding in Siberut Island, although it was not even monsoon season. The worst-affected were those living in remote areas, particularly the villages of Mongan Poula, Sri-laggai, Bojakan, and Sotboyak, where flood waters filled houses, farm huts, and boats without much warning. People immediately fled to higher ground waiting for external help, which came quickly to some badly hit areas. Many houses were partially or fully submerged by water; others were completely washed away. The flood waters also caused extensive damage to crops, livestock, other household assets, and infrastructure, especially bridges and roads.

We visited some of the worst-affected villages to interview people affected by floods. A Mentawai man told us that annual floods or localized floods were more or less predictable. However, he said, a great flood over the whole island of Siberut was unprecedented. According to the locals in Mongan Poula and Sotboyak, they were frequently hit by floods, particularly during and after logging activities of *Koperasi Andalas Madani* (HPH) and PT. *Alam Indah Lestari* (IPK) that cleared the forests in the upstream areas of the watershed of the Sikabaluan river system. Floods increased due to environmental damage (deforestation) in upstream areas.

Some residents of a remote village in South Siberut saw that a boat carrying some government healthcare workers came to the village after the flood. When they saw the village clinic submerged, they left without surveying the area. We were in the house of an elderly couple. The old lady asked for *obat sakit kepala* (painkillers). She said she was giddy and had not eaten much since the floods had prevented her husband from going to their farm to look for some food. The failure of disaster relief efforts to recognize the pressing needs of many families affected by the floods reflected the poor delivery of public services to people in remote areas.

In Muara Siberut, South Siberut sub district administration centre, we were walking through a densely populated coastal settlement and noticed that high tides had covered the road and waves were nearly lapping at the houses. Only a small patch of sandy beach prevented the waves from coming in. There had once been many coconut palms on the beach. Most had been washed away and those remaining only bore rotten roots and lower trunks. Without the trees to shield coastal settlements against strong storms and waves and absorb run-off water, heavily populated areas are now more exposed to the effects of natural disasters.

When the flood waters receded and the government lifted the emergency status, many refugees returned to their villages. In two villages in Central Siberut, men were seen breaking down the houses destroyed in the floods. They claimed they had been given materials to rebuild the house. We found out that only families with severely damaged homes had been compensated with some cash and emergency food provisions. These were given to the headman—all Mentawai village heads are men—to distribute to such families. As men were recognized as the traditional heads of households, the money was given to them. Men also played an active role in building houses and repairing damaged village roads, bridges, etc., as part of post-disaster reconstruction. These community tasks came with cash and free meals.

Women, children, and other men not involved would gather to watch those at work, in what was traditionally a show of community spirit. In earlier times, the Mentawaians had their own methods and traditional knowledge to build the *uma,* involving special rites performed by elderly women and men. Local materials such as wood, palm leaves, and bamboo were used, each with its unique technical function and cultural and symbolic meaning. The entire family or community, and sometimes even adjacent villages, were involved in the construction. Today, houses are built using coral sand (which degrade the coral reefs) or cement and zinc sheets. We saw Mentawaians carrying bags of cement and other commodities on the ship from mainland Padang to the islands.

However, there is a recognition of such gendered vulnerability in land ownership among the Mentawaians themselves, and the skewed gender ownership patterns are questioned even by men themselves. Several men acknowledged the many difficulties women and girls face in disaster situations, after several questions to them about disaster response and women's and men's experiences, including an awareness that providing women with access to land and land ownership is important. Without such access, women's preparedness and response to disasters is woefully inadequate. A man from Central Siberut said, "If according to cultural norms, I would pass all my land to my son to look after it when I am old or when I die. But I willed my lands to be shared with my son and two daughters. It is really a big problem if my daughters do not have any means of future livelihood for survival."

7.6 Conclusion

Mentawaians depend greatly on natural resources, forests, land, rivers, and their surroundings for their survival, and many of these resources are fragile and can be harmed by natural disasters or destructive human activities. The living conditions and the environment of the indigenous Mentawaians are harsh and lack many basic facilities, so the people do not have readily available resources to adapt or cope after a disaster. The inaccessibility of remote areas of the islands makes disaster rehabilitation and recovery more difficult. Yet, for generations, the Mentawaians have responded appropriately to disasters based on ancestral wisdom, knowledge, skills, and experiences:

- They lived in close contact with nature and had deep knowledge—especially the *sikerei* and older women and men—about the value, benefits, and functions of these resources in maintaining harmony between humans and nature.

- They could interpret signs in nature to recognize dangers and flee to safer, higher ground.
- They relied on animals that can sense disturbances in nature.
- They were protected from natural disasters by dense forests and coral reefs, but these natural resources are being destroyed for profit, increasing the vulnerability of the land and the people.

Mentawaians have realized that guaranteed rights to lands for subsistence and food security would help reduce disaster vulnerability. For this, they have been strengthening their bargaining power and rights as indigenous people through capacity building by working with NGOs (e.g., YCM) or through political participation at local and village levels. Another community initiative involves peaceful action such as protests, petitions, and dialogues with the local government and decision makers to oppose logging, oil palm plantations, and other destructive activities on their islands. Such initiatives are supported by local NGOs such as YCM and *Wahana Lingkungan Hidup Indonesia* (WALHI). Some such actions have been fairly successful in influencing government decisions and exerting pressure to reduce deforestation and the vulnerability of the islands to natural disasters.[9]

However, gender issues need greater attention. Much of the post-disaster assistance was based on land ownership, and women, without any ownership to land, did not have direct access to assistance. They could only receive support through their male members of the household. However, it was also noted that women were excluded not only because of their weak claim to land, but also because they had weaker access to information, which in turn was the result of their lack of mobility and lack of access to transportation. If they had access to information, they could have received assistance even without land ownership. However, since many women did not have either land or access to information, they were doubly disadvantaged.

This study clearly shows that the relationship between gender, land tenure, and disasters is complex, but remains relatively unexamined. Vigorous documentation and analyses aimed at the creation of a database where the Mentawaians contribute as primary researchers and knowledge experts can fill this gap. There is still the need to be conscious that Mentawai women's access to land rights and family resources is limited in Mentawai patrilineal society. However, some men also problematize this inequality and we have seen cases where men are beginning to allocate lands equitably to sons and daughters. An understanding of gendered land tenure is crucial to grasp how women and men are affected by disaster, and needs to be taken into consideration in both pre- and post-disaster support.

---

[9]"Struggle against logging goes on in Mentawai," in: *Down to Earth*, No. 68 (February 2006); at: http://www.downtoearth-indonesia.org/story/struggle-against-logging-goes-mentawai.

References

Bappeda and BPS, 2011: *Badan Perencanaan Pembangunan Daerah* (Bappeda) *dan Biro Pusat Statistik* (BPS) *Kabupaten Kepulauan Mentawai. Kabupaten Kepulauan Mentawai dalam Angka 2011* [Development Planning Agency and Central Board of Statistics Mentawai Island Regency. Mentawai Island Regency in Figure 2011] (Jakarta: Bapedda and BPS Kepulauan Mentawai, Tuapeijat).

Barber, Charles V.; Afif, Suraya; Purnomo, Agus, 1997: *Meluruskan Arah Pelestarian Keanekaragaman Hayati dan Pembangunan di Indonesia* [Preservation of Biodiversity and Sustainable Development in Indonesia] (Jakarta: Yayasan Obor).

Cupples, Julie, 2007: "Gender and Hurricane Mitch: Reconstructing Subjectivities after Disaster", in: *Disasters*, 31,2: 155–175.

Darmanto; Setyowati, Abidah B., 2012: *Berebut Hutan Siberut: Orang Mentawai, Kekuasaan, dan Politik Ekologi* [Siberut Forests Under Seige: The Mentawaians, Power and Political Ecology.] (Jakarta: Kepustakaan Populer Gramedia).

Enarson, Elaine, 1998: "Through Women's Eyes: A Gendered Research Agenda for Disaster Social Science", in: *Disasters*, 22,2: 157–173.

Fordham, Maureen F., 1998: "Making Women Visible in Disasters: Problematizing the Private Domain", in: *Disasters*, 22,2: 126–143.

Fulu, Emma, 2007: "Gender, Vulnerability, and the Experts: Responding to the Maldives Tsunami", in: *Development and Change*, 38,5: 843–867.

Hernawati, Tarida S., 2004: *Saureinu' Sesuatu Yang Hilang* [Saureinu' Vanishing Traditions] (Padang: Yayasan Citra Mandiri [YCM]).

Hernawati, Tarida S., 2007: *Uma: Fenomena Keterkaitan Manusia dengan Alam* [Uma: The Link Between the Phenomenon of Human and Nature] (Padang: Yayasan Citra Mandiri [YCM]).

Office of Tourism, Arts & Culture of Mentawai, 2004: Tourism Development Master Plan of Mentawai District, Final Draft Report for Office of Tourism, Arts & Culture of Mentawai. Prepared by PT. Modulatama Intikreasi Consultant, Padang [Original in Indonesian: *Pemerintahan Kabupaten Kepulauan Mentawai Dinas Pariwisata Seni dan Budaya*] (2004). [*Rencana Induk Pengembangan Pariwisata Daerah (RIPPDA) Kabupaten Kepulauan Mentawai*]. Laporan Draft Final (Padang: PT. Modulatama Intikreasi Consultant).

Puailiggoubat, 2013: *Tanah Huntara Menuai Konflik* [Lands for Temporary Housing Result in Conflict] 1 March; at: http://www.puailiggoubat.com/index.php?mod=berita&id=2198.

Siahaan, Frans, 2009: "Forest Decentralization Policy in Siberut Island: The Impacts on Policy Making, Government Finance, and Local Community" (Padang: YCM [(unpublished]).

Sieh, Kerry, et al., 2008: "Earthquake Supercycles Inferred from Sea-Level Changes Recorded in the Corals of West Sumatra", in: *Science*, 322, 5908 (12 December): 1674–1678.

UNESCO, 2004: "Siberut Biosphere Reserve, Indonesia"; at: http://www.unesco.or.id/activities/inter/199.php (8 October 2004).

UNESCO, 2005: "Siberut Biosphere Reserve Activities: A New Partnership"; at: http://www.unesco.or.id/activities/science/env_sci/sitsup_env/85.php# (4 October 2005, 18 March 2007).

WWF, undated: Mentawai Islands rain forests (IM0127); at: http://www.worldwildlife.org/wildworld/profiles/terrestrial/im/im0127_full.html (18 March 2007).

YCM, 2006: *Investigasi Dan Kampanye Illegal Logging, Korupsi Dan Kejahatan Sektor Kehutanan Yang Dilakukan Oleh Perusahaan Kayu (IPK) Di Kabupaten Kepulauan Mentawai* [Investigation and campaign against Illegal Logging, Corruption and Forest Crime Conducted by Small Scale Logging Compan ies (IPK) in Mentawai Island Regency]. An internal report written by Frans Siahaan for Yayasan Citra Mandiri (YCM), Padang.

YCM, 2007: *Mentawai: Small Islands, Deforestation, Cultural Identity, and the Indigenous People* (Padang: Yayasan Citra Mandiri [YCM]).

YCM; AMA-PM Siberut Utara; WALHI Sumbar, 2004: *Laporan Investigasi Lapangan: Fakta Kerusakan Lingkungan dan Dampak Sosial Akibat Operasionalisasi HPH dan IPK di Pulau Siberut Kabupaten Kepulauan Mentawai* [Field Investigation Report: Facts on environmental destruction and social impact of HPH and IPK in Siberut Island of Mentawai Islands Regency] (Padang: Yayasan Citra Mandiri [YCM]).

About the Editors

Kyoko Kusakabe (Japan/Thailand) is an Associate Professor of Gender and Development Studies in Asian Institute of Technology. She teaches post-graduate courses, carries out research and outreach activities in the areas of gender, work, mobility and development in various countries in Asia. Her recent research is on gender issues in labour migration, cross-border agriculture investment and border trade. Her publication include: *Thailand's Hidden Workforce: Burmese Women Factory Workers* (London: Zed Books, 2012); (with Ruth Pearson): *Transformations of indigenous people's lives and livelihoods: Mobilities in China, Laos, and India,* (London: Routledge, 2013); (co-edited with Ragnhild Lund, Smita Mishra Panda and Wang Yunxian): *Gender, road and mobility in Asia* (London: Practical Action Publishing, 2012); (with Ruth Pearson): "Cross-border childcare strategies of Burmese migrant workers in Thailand", in: *Gender, place and culture,* 20,8: 960–978 (2013).

Address: Assoc. Prof. Kyoko Kusakabe, Gender and Development Studies, School of Environment, Resources and Development, Asian Institute of Technology, Klong Luang, Pathum Thani, Thailand 12120
Email: kyokok@ait.ac.th
Website: http://www.asdu.ait.ac.th/faculty/FacultyByID.cfm?FacultyID=414.

Rajendra P. Shrestha (Nepal) is an Associate Professor and Coordinator of Natural Resources Management at the Asian Institute of Technology, Thailand. His research interests include human dimension of land use/cover change including land tenure issues, land degradation, natural resources management and Geoinformatics application. He has authored more than fifty articles in internationally referred journals, twelve book chapters and number of workshop proceedings in his area of research interests. He also has co-edited two books on *Atmospheric*

K. Kusakabe et al. (eds.), *Gender and Land Tenure in the Context of Disaster in Asia*, SpringerBriefs in Environment, Security, Development and Peace, DOI 10.1007/978-3-319-16616-2

Brown Cloud (ABC); (with R.M. Shrestha, N.T. Kim Oanh, M. Rupakheti, S. Permadi, T. Kanabkaew, and M. Iyngararasan): *Emission Inventory Manual* (Nairobi: United Nations Environment Programme, 2013); (with Trisurat, Y. and R. Alkamade): *Land Use, Climate Change and Biodiversity Modeling: Perspective and Applications* (USA: IGI Publishers, 2011).

Address: Associate Professor, Natural Resources Management, School of Environment, Resources and Development, PO Box 4, Klong Luang, Pathum Thani, Thailand 12120
Email: rajendra@ait.ac.th
Website: http://www.asdu.ait.ac.th/faculty/FacultyByID.cfm?FacultyID=424.

Veena N. (India) was a journalist in India, Tanzania and Thailand before moving into academia. She is a freelance researcher and writer with interest in the fields of gender, indigenous culture and development. Her publications include: (co-authored with Anitha Venkatesh): "Mobility patterns and gendered practices among Soliga people in Karnataka, India", in: *Gender, Mobilities, and Livelihood Transformations: Comparing Indigenous People in China, India, and Laos* (London: Routledge, 2013); (co-authored with Govind Kelkar and Girija Shrestha): "Women's Agency and IT Industry in India", in: *Gender and the Digital Economy: Perspectives in the Developing World* (New Delhi: Sage, 2005); (co-authored with Kyoko Kusakabe): "Women's Work, Men's Fantasy: The Internet and Sex Work in Thailand", in: *The Gender Politics of ICT* (London: Middlesex University Press, 2005).

Address: Veena N., Gender and Development Studies, School of Environment, Resources and Development, Asian Institute of Technology, Klong Luang, Pathum Thani, Thailand 12120
Email: veena.n@ait.ac.th.

About the Authors

Bhuvaneswari Raman (India) studied at the School of Planning and Architecture in Chennai, Asian Institute of Technology, Bangkok and at the London School of Economics, UK. Her publications include "The Rhetoric and Reality of Transparency Initiatives: Transparent Cities, Opaque Power and Urban Poor's Claims to Land", in: *Journal of Community Informatics*, 8,2 (2012): 866–909; "Illegible Claims, Legal Titles, and the Worlding of Bangalore", in: *Revue Tiers Monde*, 206 (April–June 2011); (with Bawa, Z.): "Interacting with the State via Information and Communication Technologies. The Case of Nemmadi Kendras in Karnataka", in: *Media Asia, an Asian Mass Communication Quarterly*, 38,1 (2011): 52–62.

Address: Associate Professor Bhuvaneswari Raman, Jindal School of Government and Public Policy, Jindal Global University, Sonepat 131001, India
Email: raman.bhuvaneswari@gmail.com

Andreas Burghofer (Austria) is an activist of human rights and environment non-governmental organizations of long-standing experience, with an engineering background. Over the past two decades, he has been actively involved with the journal *INDABA* on Southern Africa. He has also been involved for many years with the magazine *bedrohte völker* [threatened people], which focuses on minorities and endangered people. He has published numerous articles and book reviews in magazines and German-speaking media on issues of environment, biodiversity, minority rights and politics. His books included *Nachhaltigkeit in Namibia* [Sustainability in Namibia]; *Das Schatztruhe Regenwald-Buch* [Book on Treasures of Rainforests]; *Der tropische Regenwald* [The Tropical Rainforests]; and he also translated and edited *Brahmas Haar—Die Mythologie der indischen Pflanzenwelt* (original in English Brahma's Hair—Myths of the Indian Plant Kingdom by Maneka Gandhi). He is co-author/contributor of the report: *Destruction at dawn: The rights of indigenous peoples in the Republic of Nepal.*

Address: Andreas Burghofer, Untere Viaduktgasse 53/7a, A-1030 Wien
Email: andreas.burghofer@gfbv.at
Website: www.gfbv.at.

K. Kusakabe et al. (eds.), *Gender and Land Tenure in the Context of Disaster in Asia*, SpringerBriefs in Environment, Security, Development and Peace, DOI 10.1007/978-3-319-16616-2

Subhangi M.K. Herath (Sri Lanka) is a Senior Lecturer in sociology and the former Head of the Department of Sociology at the University of Colombo, Sri Lanka. She received her PhD from the University of Waterloo, Canada in 1997. Her publications include: "Fighting for Rights or Having Fun?"—A case Study of Political awareness and Participation of Rural Sinhalese Women", in: Selvi Thiruchandran (Ed.), 1999: *Women Narration and Nation: Collective Images and Multiple Identities* (Delhi: Vikas Publishers); (with Bogdan Lešnik): "Internally displaced persons: Implications for the development of social work in Sri Lanka", in: Shula Ramon (Ed.), 2008: *Social work in the context of political conflict* (London: Venture Press): 77–97; "Care for the Elderly in Sri Lanka", in: *Socialno Delo,* 47 (June–December 2008) (Ljubljana, Slovenia, University of Ljubljana, Faculty of Social Work); (with Darja Zaviršek): "'I want to have my future, I have a Dialogue': Social work in Sri Lanka between neo-capitalism and human rights", in: *Social Work Education*, 29,8 (2010): 831–842.

Address: Senior lecturer, Department of Sociology, University of Colombo, Colombo 3, Sri Lanka
Email: subhangiherath40@yahoo.com.

Ramanie Jayatilaka (Sri Lanka) is an Associate Professor in Sociology at the University of Colombo. She is a Board Member of the *Centre for Women's Research* (CENWOR) and Board Member of *Sri Lanka Centre for Development Facilitation* (SLCDF). Her research interests are in the areas of gender issues, international female migration, participatory rural development and social aspects of health. Two of her publications include: *Women Migrant Workers & Trafficking in Sri Lanka* (Colombo: Women's Education and Research Centre, 2008), and *Globalization and Rural Women: A Case Study of Two Villages in the District of Galle* (Colombo: Centre for society and religion, 1998).

Address: Associate Professor Ramanie Jayatilaka, Department of Sociology, University of Colombo, Colombo 3, Sri Lanka
Email: ramaniej@yahoo.com.

Kyoko Kusakabe (Japan/Thailand) is an Associate Professor of Gender and Development Studies in Asian Institute of Technology. She teaches post-graduate courses, carries out research and outreach activities in the areas of gender, work, mobility and development in various countries in Asia. Her recent research is on gender issues in labour migration, cross-border agriculture investment and border trade. Her publication include: *Thailand's Hidden Workforce: Burmese Women Factory Workers* (London: Zed Books, 2012); (with Ruth Pearson): *Transformations of indigenous people's lives and livelihoods: Mobilities in China, Laos, and India,* (London: Routledge, 2013); (co-edited with Ragnhild Lund, Smita Mishra Panda and Wang Yunxian): *Gender, road and mobility in Asia* (London: Practical Action Publishing, 2012); (with Ruth Pearson): "Cross-border childcare strategies of Burmese migrant workers in Thailand", in: *Gender, place and culture,* 20,8: 960–978 (2013).

Address: Assoc. Prof. Kyoko Kusakabe, Gender and Development Studies, School of Environment, Resources and Development, Asian Institute of Technology, Klong Luang, Pathum Thani, Thailand 12120
Email: kyokok@ait.ac.th
Website: http://www.asdu.ait.ac.th/faculty/FacultyByID.cfm?FacultyID=414.

Mahdi (Indonesia) is a chair of Integrated Natural Resources Management field of Study at Graduate Programme of Andalas University in Padang, Indonesia. He obtained his PhD in Natural Resources Management from the *Asian Institute of Technology* (AIT) Thailand. His research interests include natural resource and economic aspects of natural and environmental resources.

Address: Mahdi, Komplek Cemara I Blok BB No. 3 Gunung Pangilun, Padang 25264, Indonesia
Email: kuteihmahdi@gmail.com.

Frans R. Siahaan (Indonesia) is an environmental justice activist with extensive experiences in community organizing, indigenous peoples' rights, and natural resources governance. He obtained his Masters in Public Administration from the University of Massachusetts, USA. He has worked for almost 11 years with *Yayasan Citra Mandiri* (YCM), an NGO that works with the Mentawai people. He is co-author (with Fajri Nailus and Denni Nurdwiansyah) of *Mining policy in West Kalimantan* (2015). He has also written widely on issues of Mentawai and environmental governance, published in local and national media in Indonesia. He has also contributed to book chapters, edited books, articles and YCM research reports, including "Mentawai at Glance" (Chap. 1, in *Uma: The link between human and nature*) and "Forest decentralization policy in Siberut Island: The impacts on policy making, government finance, and local community".

Address: Frans Siahaan, Program Officer–Environmental Governance at The Asia Foundation, Jakarta, Indonesia.
Email: frans.siahaan@gmail.com.

Veena N. (India) was a journalist in India, Tanzania and Thailand before moving into academia as a researcher and writer with interest in the fields of gender, indigenous culture and development. Her publications include: (co-authored with Anitha Venkatesh): "Mobility patterns and gendered practices among Soliga people in Karnataka, India", in: *Gender, Mobilities, and Livelihood Transformations: Comparing Indigenous People in China, India, and Laos* (London: Routledge, 2013); (co-authored with Govind Kelkar and Girija Shrestha): "Women's Agency and IT Industry in India", in: *Gender and the Digital Economy: Perspectives in the Developing World* (New Delhi: Sage, 2005); (co-authored with Kyoko Kusakabe): "Women's Work, Men's Fantasy: The Internet and Sex Work in Thailand", in: *The Gender Politics of ICT* (London: Middlesex University Press, 2005).

Address: Gender and Development Studies, School of Environment, Resources and Development, Asian Institute of Technology, Klong Luang, Pathum Thani, Thailand 12120
Email: veena.n@ait.ac.th.

Wee Aik Pang (Malaysia) is a Sarawakian who has spent the last two decades learning and engaging with mostly indigenous communities of his home state while making sense of the anthropological and sociological imaginations of societies in general but communities in specific. He continues to work with and challenges some rural communities on development discourses while trying to comprehend the people's worldviews. He also conducts independent research and writes on community challenges and related issues. He is the author of *Guidebook on reclaiming Sarawak Native Customary Rights (NCR) lands in courts* and co-wrote (with three others) *Land is life: Land rights and oil palm development in Sarawak.*

Address: Wee Aik Pang, No 9, 1st Floor, Maong Bazaar, 3rd Mile Rock Road, 93250 Kuching, Sarawak, Malaysia
Email: weeaikpang@gmail.com.

Yonariza (Indonesia) is a professor in forest resources management at Andalas University, Padang. Indonesia. He has published book chapters, journal articles, discussion papers, and conference proceedings. His topics cover local forest management policy, the rights of indigenous people, post-logging ban timber tree planting in Thailand and the Philippines, relationship between protected area and local livelihood, irrigation management, decentralization policy and revitalization of local institutions for protected area co-management, rural household participation in timber felling in a protected area, and Southeast Asia land and resources tenure. He received research grants from the Toyota Foundation, Nippon Foundation, and Ford Foundation as well as from the government of Indonesia.

Address: Prof. Yonariza, Komplek Unand DII/05/01, Kel. Banda Buek, Kec. Lubuk Kilangan, Padang 25231, Indonesia
Email: yonariza@hotmail.com
Website: http://yonariza.com/.

Carol Yong (Malaysia) is a feminist advocate activist with many years of experience in researching, writing, community training and interpreting-translating, with a focus on the rights of indigenous people, especially indigenous women in Asia, and development issues. She has acted as external consultant on evaluations of NGOs in Indonesia; gender, community-based and participatory action research; and analysing forest policies and practices for international organizations. She is author of *Flowed over: The Babagon Dam and the resettlement of the Kadazandusun in Sabah* and *Deforestation drivers and human rights in Malaysia: A national overview and two sub-regional case studies* (2014). She has also published numerous articles and book chapters on issues of land rights, development-induced displacement and resettlement, human rights, and resource politics and corruption issues, from a gender, communities' and indigenous peoples' rights perspective.

Address: Carol Yong, No. 9, 1st Floor, Maong Bazaar, 3rd Mile Rock Road, 93250 Kuching, Sarawak.
Email: yrn@vjfg.at.

About the Book

This book explores an interdisciplinary field at the intersection of gender and development studies, disaster and land tenure policy. It is well-known that women have weaker claims to land. But how does that translate to increased vulnerability during disaster? Using case studies from Asia, this book argues that land tenure is a key factor in mitigating the impact of disasters on women. The scale and frequency of disasters have been increasing in recent decades due to human impact on the landscape and climate. Unsustainable farming and land management systems have increased environmental risks and social vulnerabilities. However, the costs of disasters are disproportionately borne by women around the world. Women are more vulnerable to disaster due to reduced mobility and lack of control over assets. Post-disaster, women's vulnerabilities increase due to gendered rescue and rehabilitation practices. A gendered approach to land rights is critical to disaster preparedness and rehabilitation.

- This is a unique book focusing on gender, disaster and land tenure, especially in Asia.
- The case studies from the field stress the importance of land tenure systems during a disaster.
- This new field of interdisciplinary study introduces new areas of research and policy.
- Research focuses on India, Indonesia, Malaysia, and Sri Lanka.

List of Contents: 1. Gender and Land Tenure in the Context of Disaster: Veena N. and Kyoko Kusakabe; 2. Gender Impact of Large-scale Deforestation and Oil Palm Plantations among Indigenous Groups in Sarawak, Malaysia: Carol Yong and Wee Aik Pang; 3. Displacing Women, Resettling Families: Impact of Landslides on Women's Land Tenure Rights in Sri Lanka: Subhangi M. K. Herath; 4. Impact of Flash Floods on a Matrilineal Society in West Sumatra, Indonesia: Yonariza and Mahdi; 5. Urbanization and Disaster: Loss of Women's Property Ownership in Leh, Ladakh: Bhuvaneswari Raman; 6. A Coir Mill of Their Own: Women's Agency in

K. Kusakabe et al. (eds.), *Gender and Land Tenure in the Context of Disaster in Asia*, SpringerBriefs in Environment, Security, Development and Peace, DOI 10.1007/978-3-319-16616-2

Post-tsunami Sri Lanka: Ramanie Jayatilaka; 7. Gender, Land Tenure, and Disasters in the Mentawai Islands, Indonesia: Carol Yong, Frans R. Siahaan, and Andreas Burghofer

More on this book is at:
http://www.afes-press-books.de/html/SpringerBriefs_ESDP_21.htm.

MIX
Papier aus verantwortungsvollen Quellen
Paper from responsible sources
FSC® C105338

FSC
www.fsc.org

If you have any concerns about our products,
you can contact us on
ProductSafety@springernature.com

In case Publisher is established outside the EU,
the EU authorized representative is:
Springer Nature Customer Service Center GmbH
Europaplatz 3, 69115 Heidelberg, Germany

Printed by Libri Plureos GmbH
in Hamburg, Germany